GeoAI

GeoAI

Artificial Intelligence in GIS

EDITED BY

Ismael Chivite
Nicholas Giner
Matt Artz

Esri Press
Redlands, California

Esri Press, 380 New York Street, Redlands, California 92373-8100
Copyright © 2025 Esri
All rights reserved.
Printed in the United States of America.

29 28 27 26 25   1 2 3 4 5 6 7 8 9 10

ISBN: 9781589488441

The information contained in this document is the exclusive property of Esri or its licensors. This work is protected under United States copyright law and other international copyright treaties and conventions. No part of this work may be used, used to train, or used to prepare derivative works, copied, reproduced, redistributed, publicly displayed or performed, or transmitted in any form or by any means, electronic or mechanical, including photocopying, digital scanning, and recording, or by any machine learning, artificial intelligence, information storage, or retrieval systems, except as expressly permitted in writing by Esri. Any unauthorized use of this work to train artificial intelligence (AI) technologies is expressly prohibited. All requests should be sent to Attention: Director, Contracts and Legal Department, Esri, 380 New York Street, Redlands, California 92373-8100, USA.

The information contained in this document is subject to change without notice.

US Government Restricted/Limited Rights: Any software, documentation, and/or data delivered hereunder is subject to the terms of the License Agreement. The commercial license rights in the License Agreement strictly govern Licensee's use, reproduction, or disclosure of the software, data, and documentation. In no event shall the US Government acquire greater than RESTRICTED/LIMITED RIGHTS. At a minimum, use, duplication, or disclosure by the US Government is subject to restrictions as set forth in FAR §52.227-14 Alternates I, II, and III (DEC 2007); FAR §52.227-19(b) (DEC 2007) and/or FAR §12.211/12.212 (Commercial Technical Data/Computer Software); and DFARS §252.227-7015 (DEC 2011) (Technical Data–Commercial Items) and/or DFARS §227.7202 (Commercial Computer Software and Commercial Computer Software Documentation), as applicable. Contractor/Manufacturer is Esri, 380 New York Street, Redlands, California 92373-8100, USA.

Esri products or services referenced in this publication are trademarks, service marks, or registered marks of Esri in the United States, the European Community, or certain other jurisdictions. To learn more about Esri marks, go to: links.esri.com/EsriProductNamingGuide. Other companies and products or services mentioned herein may be trademarks, service marks, or registered marks of their respective mark owners.

For purchasing and distribution options (both domestic and international), please visit esripress.esri.com.

Contents

Introduction	ix
Part 1: GeoAI Technology Overview	**1**
Part 2: Public Sector Applications	**11**

GeoAI Helps Stave Off Pest Infestation of Hemlock Trees — 21
Forest Pest Branch, Urban Forest Management Division, Fairfax County, Virginia

Using GeoAI to Inventory ADA Curb Ramps Saves Significant Time and Money — 18
Douglas County and the City of Omaha, Nebraska

Improving Roadways Using Drone Imagery and Machine Learning — 23
Utah Department of Transportation

Drones, Maps, and AI Unite City Functions — 29
City of Vilnius, Lithuania

As Sea Level Rise Threatens, Safeguarding a Sense of Place with a Digital Twin — 34
Government of Tuvalu

Deep Learning Helps Automate Map Updates to Better Serve Citizens — 39
Government of Kuwait

GIS and AI for Precise Damage Assessments — 46
Esri

Getting the Most of GeoAI in Emergency Management — 51
Esri

Part 3: Private Sector Applications — 57

Deep Learning Model Unlocks Potential of Solar Energy Development — 58
Pivot Energy

CEOs May Be Underusing this AI Capability — 62
Bouwinvest

Another AI Capability that Business Leaders May Be Overlooking — 65
Esri

Mapping New Possibilities for Business Success — 68
Esri

GeoAI, Reality Capture, and the Future of Digital Twins — 72
Esri

Mapping the What-Ifs: Fertile Ground for AI-Powered Simulations — 77
Esri

Part 4: NGO/Nonprofit Applications — 81

Drone Mapping Helps Find Flood Victims, with AI Assistance — 82
United Nations World Food Programme

GeoAI, Corporate Responsibility, and the Vigilance of a Climate Watchdog — 91
Amazon Conservation

Mapping Land Mines and Explosive Remnants of War — 97
The HALO Trust

Part 5: Next Steps — 103

Contributors — 109

"GeoAI—the integration of spatial analysis, AI, and big data—is creating new insights that promise to transform our understanding of the world."

Jack Dangermond
Esri cofounder and president

Introduction

Organizations across the globe have long relied on geographic information system (GIS) technology to manage and analyze data through the powerful lens of location, helping them tackle some of the toughest business and societal challenges. GeoAI—GIS enriched with artificial intelligence (AI)—is already helping organizations get better, faster answers.

What is AI?

Simply stated, AI is the simulation of human intelligence in machines, training computers to recognize and detect patterns, extract information and meaning from data, and to solve problems and help make decisions through artificial learning.

Machine learning, deep learning, and generative AI are the categories of AI that are most relevant to GIS.

Machine learning is an application of AI that allows machines to learn without being specifically programmed to do so. It uncovers insights from data through methods incorporating decision trees or cluster analysis.

Deep learning is a type of machine learning based on artificial neural networks, which progressively uses multiple layers of information to extract higher-level features from the raw input. It uses more advanced methods and helps solve complex problems across large data volumes, with a focus on automated data extraction and pattern recognition.

Generative AI is a subset of deep learning that creates new data, such as text, images, videos, audio, and 3D models. Generative AI models learn patterns from existing data and use this knowledge to generate new and possibly unique outputs.

GeoAI: AI in GIS

GeoAI advances the science of GIS by using AI tools and models to automate data extraction and perform analysis on imagery, video, text, 3D, vector/tabular, time series, and other data. It extracts and classifies features from unstructured text or imagery sourced from satellites, drones, aircraft, video feeds, and even mobile phones. GeoAI is also used to detect patterns, clusters, and anomalies in data, and to make predictions and forecasts.

A key impact of GeoAI is making feature extraction and spatial analysis more widely available. This means better-informed decisions, more efficient operations, and the ability to tackle complex spatial problems that could have been previously out of reach for some organizations.

Also, organizations that already use GIS extensively will benefit from the ability to tackle complex problems by combining human GIS expertise with AI capabilities. This could lead to entirely new applications and insights that weren't possible before.

How can GeoAI help GIS professionals?

- **Automated data extraction:** AI helps GIS professionals by automating processes to extract useful GIS information from data. Object detection in aerial imagery and named-entity recognition from unstructured text are two examples. Object detection in imagery helps emergency response teams quickly map locations of debris. Named-entity recognition helps law enforcement officers process text documents in search references to events, people, and the like. Such tasks involve repetitive work; AI lets the machine do this work so that humans can focus their energy and expertise on more complex problem solving that the machine can't do. And AI does this work more quickly and at scale.

- **Deeper insights:** ArcGIS includes many tools to perform analysis of geospatial information. With AI techniques, Esri® is adding even more tools, providing GIS analysts with new options to identify patterns, make predictions, and ultimately gain better insights from data. AI brings together more powerful tools that, when used correctly, allow GIS users to do things we could not do before. Good examples include state-of-the-art machine learning tools for creating predictions using large, multivariate datasets and for making forecasts based on complex time series patterns. Multimodal analysis is also enhanced with AI, enabling analysis across unstructured text, images, and other data modalities not supported by traditional tools.

Using GeoAI

The combination of AI and GIS has already changed how leading organizations manage operations. GeoAI enables new levels of sustainability, efficiency, and growth. The next section of this book presents an overview of current GeoAI capabilities, followed by stories illustrating how organizations are already using GeoAI in the public, private, and NGO/nonprofit sectors. The book concludes with a section about the next steps you can take to learn more about getting started with GeoAI.

Learn more about GeoAI by visiting:

go.esri.com/geoai_book

Part 1
GeoAI Technology Overview

GeoAI integrates artificial intelligence, or AI, with geospatial data, science, and technology to increase understanding and solve spatial problems. AI is the ability of computers to perform tasks that typically require some level of human intelligence and reasoning—through programming that continually adapts, infers patterns, generalizes, and improves output over time. We can use GeoAI for applications such as detecting and categorizing objects in imagery and lidar, identifying clusters and anomalies in data, and making predictions and forecasts. The intersection of GIS, AI, machine learning, and deep learning creates opportunities that weren't available before.

GeoAI offers new ways to evaluate numerous solutions to difficult spatial problems. Spatially explicit models incorporate an aspect of geography, such as location, shape, or proximity, into an algorithm, making the models more efficient, accurate, and representative of the reality we want to model. With these techniques, we can allocate resources based on meaningful spatial patterns and relationships, find trends and anomalies in space and time, and incorporate spatial relationships into predictions and forecasts.

Machine learning in ArcGIS

Machine learning is a branch of AI in which computers learn patterns within data, and then use what they've learned to predict outcomes or make decisions. Machine learning algorithms are data-driven and operate with minimal human intervention. As machines process more and more data, they are trained so that they automatically "learn" how to adjust their behavior and improve their performance based on previous experience.

Machine learning shows up everywhere in our daily lives and across many industries. Product recommendations, traffic alerts, social media ads, health-care diagnoses, fraud detection, predictive maintenance—all use machine learning in some way, shape, or form. Irrespective of the specific industry or application, the types of problems solved by machine learning generally fall into three main categories: clustering, prediction (which includes regression and classification problems), and forecasting.

Machine learning in ArcGIS Pro

In the context of ArcGIS technology, machine learning is far from new. In fact, machine learning algorithms have been incorporated within ArcGIS and used in geographic applications of these three categories for many years. For example, you can classify pixels within remotely sensed data using K-Nearest Neighbor or Support Vector Machine algorithms. Or you can apply decision tree ensembles, machine learning techniques that combine multiple decision trees to improve predictive accuracy, to classification problems with vector and tabular data using the Forest-Based and Boosted Classification and Regression tool. You can also take advantage of logistic regression and maximum entropy (Presence-Only Prediction—MaxEnt) approaches for predicting binary classification outcomes.

For clustering problems, you have access to algorithms that group spatial data based solely on the data attributes (Multivariate

Clustering), their locations (Density-Based Clustering), or based on both the attributes and locations of the data (Build Balanced Zones). You also have access to a family of global regression models and decision tree ensembles for regression tasks, as well as the ability to apply decision tree ensembles to time series forecasts. Lastly, you can use Causal Inference Analysis to go beyond prediction and uncover the true causal relationships between variables. Behind the scenes, the algorithm uses machine learning to isolate the effect of a true exposure on an outcome from other confounding variables. An example would be isolating the effect of fertilizer (cause) on corn yield (effect) in the presence of other related variables, such as soil type, farming techniques, and environmental variables.

Although all these algorithms are considered machine learning, there is a fundamental difference between applying a traditional, nonspatial machine learning method to spatial data (such as, Forest-Based and Boosted Classification and Regression) and using true spatial machine learning. In the latter case, geography is incorporated directly into the mathematics of the machine learning algorithm through notions of shape, adjacency, orientation, contiguity, proximity, density, spatial distribution, and so on. Examples of spatially explicit machine learning algorithms include spatial autoregression and different types of geographically weighted regression, as well as spatially constrained multivariate clustering. Your choice of machine learning algorithm should always depend on the underlying problem you are trying to solve, the structure of your data, and your desired goals and deliverables.

AutoML

In the past decade, machine learning has experienced rapid growth in both the range of applications it is used for and the amount of new research produced. Some of the driving forces behind this growth are the maturity of the machine learning algorithms and methods, the generation and proliferation of volumes of data for the algorithms to learn from, the inexpensive computers to run the algorithms, and the increasing awareness among businesses that machine learning algorithms can address complex data structures and problems.

Many organizations want to use machine learning to take advantage of their data and derive insights, but there is an imbalance between the number of potential machine learning applications and the number of trained, expert machine learning practitioners to address them. As a result, there is an increasing demand to standardize machine learning across organizations by creating tools that make machine learning widely accessible throughout and can be used off the shelf by nonexperts in machine learning, as well as by domain experts.

Recently, automated machine learning (AutoML) has emerged as an approach to address the demand for machine learning in organizations across all experience and skill levels. AutoML aims to create a single system to automate (in other words, remove human input from) as much of the machine learning workflow as possible, including data preparation, data engineering, model selection, hyperparameter tuning, and model evaluation. In doing so, it can be beneficial to nonexperts by lowering the barrier of entry into machine learning but also to trained machine learning practitioners by eliminating some of the most tedious and time-consuming steps in the machine learning workflow.

Deep learning in ArcGIS

Deep learning is available in different formats across the ArcGIS ecosystem, making it increasingly accessible for users of varying skill sets. Whether you're interested in using ArcGIS Online to test a pretrained model or using the ArcGIS API for Python to create a custom model, there is an ArcGIS option for you. Whichever platform you choose, ArcGIS has the tools to help you accomplish your deep learning tasks.

Pretrained deep learning models

Training AI models is a time- and resource-intensive process, but ready-made pretrained GeoAI models automate the task of digitizing and extracting geographic features from imagery, point cloud, and text datasets.

Manually extracting features from raw data, such as digitizing building footprints or generating land-cover maps, is time consuming. Deep learning automates the process and minimizes the manual interaction necessary to complete these tasks. However, training a deep learning model can be complicated because it requires large quantities of data, computing resources, and knowledge of how deep learning works.

With pretrained models, analysts do not need to invest time and effort in training a deep learning model. The models have been trained on data from a variety of geographies. As new imagery becomes available, we can extract features and produce layers of GIS datasets for mapping, visualization, and analysis. Pretrained models can be accessed from ArcGIS Living Atlas of the World and other online repositories.

More than 100 pretrained models are already available, and even more are being developed every day, including the following:

- **Image feature extraction and detection:** extract features, such as buildings, vehicles, swimming pools, and solar panels, from aerial and satellite imagery.
- **Pixel classification:** classify land-cover satellite imagery.
- **Point cloud classification:** classify power lines and tree points using point cloud data.
- **Image redaction:** blur sensitive areas from imagery to comply with privacy policies.

- **Object tracking:** track moving objects, such as vehicles, in motion imagery.
- **Named-entity recognition:** identify or categorize entities from text.

Additionally, foundational models such as Prithvi have been trained on geospatial data, and vision-language models (such as OpenAI's GPT-4 and GPT-4o, as well as Llama) have been integrated and now bring generative AI capabilities to GeoAI.

Deep learning geoprocessing tools and wizards in ArcGIS Pro

ArcGIS Pro is a desktop application that includes deep learning tools with the ArcGIS Image Analyst, ArcGIS Spatial Analyst™, and ArcGIS 3D Analyst™ extensions, as well as in the GeoAI toolbox.

ArcGIS Pro has deep learning capabilities within a suite of geoprocessing tools. This familiar environment provides an intuitive workflow for existing ArcGIS Pro users while also providing ample user-friendly wizards for collecting training data and configurable parameters for tuning your models.

ArcGIS Pro has a wizard to guide learning and geoprocessing tools for all steps in the imagery-based deep learning workflow, including the Label Objects for Deep Learning pane and Training Samples Manager for training sample collection. It includes the Train Deep Learning Model tool to train a model and task-specific geoprocessing tools for model inferencing, including Detect Objects, Classify Objects, and Classify Pixels.

The GeoAI toolbox in ArcGIS Pro contains tools for using and training AI models that work with geospatial and tabular data. These tools use modern machine learning and deep learning techniques and integrate them with GIS. The GeoAI toolbox contains tools that allow you to train and use models that perform classification and regression on feature and tabular datasets, as well as models that classify, transform, and extract information from unstructured text using natural language processing (NLP). The tools in the Feature and Tabular Analysis toolset use AutoML to train and fine-tune, as well as create ensembles of, the best machine learning models using the data and available computer resources. The trained models can be used for predicting both categorical variables (classification) and continuous variables (regression) on similar datasets. The tools in the Text Analysis toolset allow you to use and fine-tune pretrained text and NLP models from ArcGIS Living Atlas or create models using labeled text data. The tools in this toolset also work with models created using the ArcGIS API for Python arcgis.learn module. The models created by these tools can be used in and further fine-tuned using the Python API.

Map Viewer analysis tools in ArcGIS Online

ArcGIS Online is a cloud-based software as a service (SaaS) that includes some deep learning tools in Map Viewer.

Are you just looking to try out a pretrained deep learning model from a colleague or ArcGIS Living Atlas? The deep learning analysis tools in ArcGIS Online may be the best place for you to get started with deep learning in ArcGIS. The three analysis tools, Detect Objects Using Deep Learning, Classify Objects Using Deep Learning, and Classify Pixels Using Deep Learning, are easy to use and allow you to use input imagery and models from your organization or shared through ArcGIS Online and ArcGIS Living Atlas.

The analysis tools in Map Viewer allow you to use a pretrained model to gain inference from data, but they do not provide the same model training capabilities as the other platforms, such as ArcGIS Pro. This makes ArcGIS Online a less powerful platform for using deep learning but also significantly more user-friendly with a lower barrier to entry than the other options. If your aim is to use an existing model on data available in ArcGIS Online, the deep learning analysis tools in Map Viewer are the simplest and most intuitive way to do that.

Deep Learning Studio in ArcGIS Enterprise

Deep Learning Studio is a web app available with ArcGIS Image Server for ArcGIS Enterprise. It is a comprehensive and collaborative application for model training and inferencing.

Deep Learning Studio includes multiple tools to manage the workload of building and using a deep learning model across multiple users. The first step in using Deep Learning Studio is to create a project, and then you can set the data source, set the scheme for training samples, invite project members, and set up work units. The primary advantage of Deep Learning Studio is that it enables collaboration by designating project members and dividing the project into discrete work units.

Project members can be chosen from groups in your ArcGIS Enterprise Organization. A contributor's capabilities are divided into three roles—project owner, analyst, and sample collector—and are related to their user privileges in the ArcGIS Enterprise Organization—for example, anyone with editing privileges can have the sample collector role.

Work units are sections of the imagery that are split up so that individual users can collect training data in an organized fashion, without overlap. Work units can be set up in three ways:

- **Grid system:** You can define the size of the grid that overlays the imagery so that each member works on one grid cell at a time.
- **Custom work units:** You can use an existing polygon feature layer to define polygonal areas for each member to work on.
- **Individual images:** For imagery with multiple individual images, each member can work on one image at a time.

The designation of project members and division into work units allows project members to collaborate on collecting and reviewing training samples. Once an area is marked complete, project analysts and owners can review the work.

Can you still collaborate with other platforms? Sure. As with other analysis workflows, data can always be shared at intermediate steps. For example, if you wanted to have multiple users collect training data, you could do that in separate instances of ArcGIS Pro and merge the feature classes together to create one large training dataset. However, there is no built-in way of performing this collaboration except in Deep Learning Studio, which makes it an ideal platform to perform deep learning when multiple people are involved.

ArcGIS API for Python

Understanding Python unlocks more capabilities in any GIS workflow, and deep learning is no exception.

The Python API is a Python library for scripting workflows across the ArcGIS suite, including for GIS organization administration, content management, and spatial analysis/data science. Its functionality includes the arcgis.learn module, with various functions for manipulating data and training a deep learning model. The Python API is free to install but most of its functionality requires it to be connected to an ArcGIS account (ArcGIS Developer/Platform account, ArcGIS Online account, or ArcGIS Enterprise account). The type of account, user type, and role you need depends on the resource you want to access and the operations you want to undertake.

The Python API currently offers flexibility in model training that is unique in the ArcGIS ecosystem. These tools are available only in Python, so with no user interface, you must have working experience in Python to use them successfully. The Python API arcgis.learn module offers options to incorporate external models into the ArcGIS workflow, as well as more options to tune hyperparameters and augment your data. Using the Python API library allows users to leverage some of the best open-source machine learning and deep learning libraries and seamlessly integrate them into the ArcGIS ecosystem.

Another benefit of using the Python API is that it is designed for web-hosted features so you can work with hosted imagery and feature classes.

However, with no user interface, collecting training data must be done outside the Python workflow. You can use the tools in ArcGIS Pro or other tools to collect training data as a feature class and then use the export_training_data function to convert the training data to the correct format and continue the work in Python.

Programmers also have other options without using the Python API. All the ArcGIS Pro geoprocessing tools, including the deep learning tools, are available with the ArcPy library. You can still train your model with ArcPy, but it offers less flexibility and fewer advanced options than the Python API. Python users can also combine the ArcPy and Python API libraries with other open-source libraries to maximize flexibility and derive the full benefits of each library.

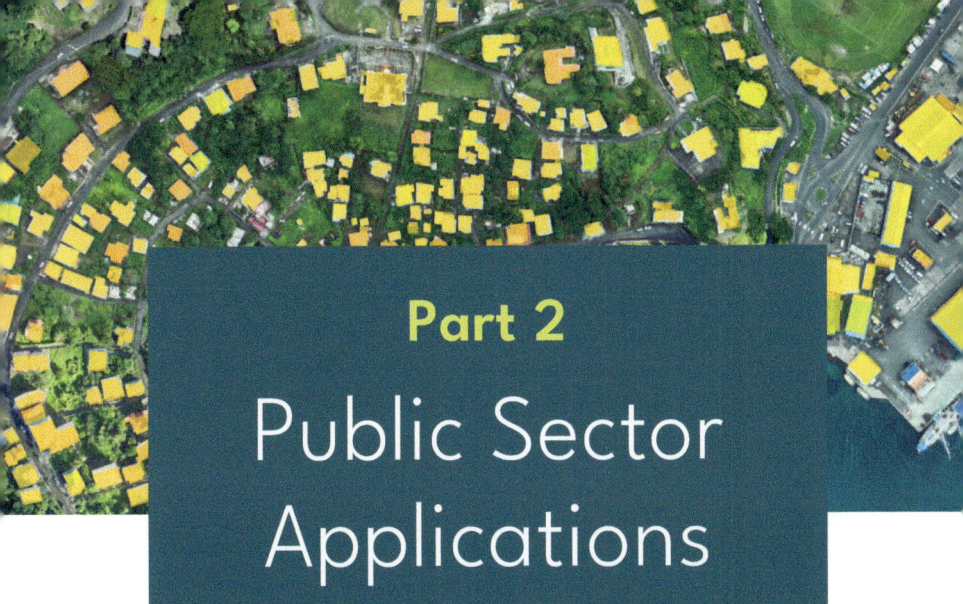

Part 2
Public Sector Applications

In the public sector, planners, decision-makers, and GIS experts can collaborate using GeoAI to answer critical questions faster and with greater confidence. For example, in the case of disaster response, officials can use GeoAI to quickly assess road damage for emergency response services or prioritize where to perform infrastructure maintenance and provide other aid for residents impacted by the disaster. In addition, government agencies can use advanced spatial analysis for policy decisions, resource allocation, and public services.

GIS in action

This section presents real-life stories about how GeoAI helps public-sector organizations save time and money, improve services, and more.

GeoAI Helps Stave Off Pest Infestation of Hemlock Trees

Forest Pest Branch, Urban Forest Management Division, Fairfax County, Virginia

In the large, urbanized area southwest of Washington, DC, residents of Fairfax County, Virginia, enjoy vast canopies of evergreen trees—including hemlocks—year-round. But there is currently a pest devastating hemlock trees throughout the southeastern United States. The hemlock woolly adelgid feeds on the sap and water storage cells at the base of a tree's needles, where the pests also lay eggs, causing an infestation. The "woolly" part of the pest's name comes from the fluffy white appearance of the insect's egg masses.

The Forest Pest Branch of the county's Urban Forest Management Division (UFMD) is responsible for monitoring a variety of forest insect pests and tree diseases; it also provides public outreach and education to help residents take part in tree preservation efforts. Staff at the Forest Pest Branch understood that managing the hemlock woolly adelgid infestation and preserving Fairfax County's existing contiguous tree canopy were critical.

Mapping plays an important role in these endeavors. UFMD staff needed to inventory the hemlocks and map their locations as part of the management plan, but searching through aerial photography of thousands of acres to find them was going to be too time-consuming and labor-intensive. They needed a way to quickly prioritize

The hemlock woolly adelgid lays eggs at the base of the hemlock tree's needles.

field visits to areas that were likely to have large numbers of hemlock trees. From there, they could target areas for protective measures against infestation.

To do this, in 2023, UFMD urban forester Patrick O'Brien enlisted the county's GIS Division. Fairfax County GIS analyst Greg Bacon decided to use ArcGIS technology and GeoAI to automate the search for evergreens in aerial photography in hopes of finding the hemlocks among them.

Automating tree detection with a deep learning model

Using the analytical power of ArcGIS Pro alongside ArcGIS Image Analyst, Bacon created a unique workflow to find the hemlocks.

"UFMD is a longtime user of the county's aerial photography. They've used the data in desktop and mobile mapping but have also worked with the GIS Division on more advanced projects such as land-cover classification," said Bacon. "This project continued our

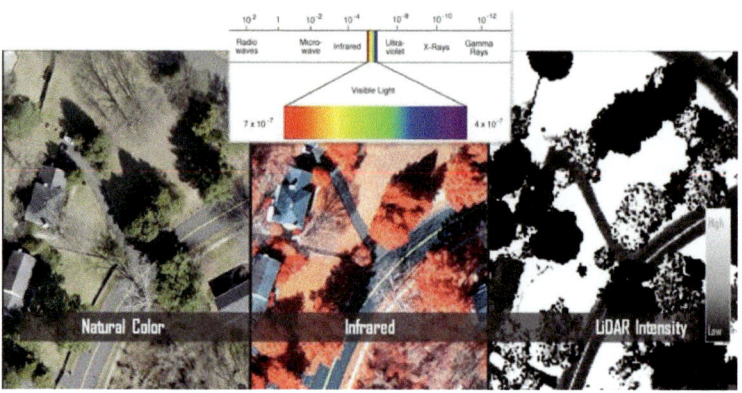

The team filtered the tree data by analyzing the greenness of pixels, infrared information, and lidar intensity values within the imagery.

collaboration in finding new ways for aerial photography to enhance their work and address evolving needs."

Bacon selected leaf-off aerial orthophotography from 2022 as the input for a tree-detection deep learning model that is currently available as a pretrained model in ArcGIS Living Atlas of the World. ArcGIS pretrained models automate the process of digitizing and extracting geographic features from imagery and point cloud datasets.

This model was a great starting point for creating tree detection geometries on a variety of supported imagery. The model produced good results—yet it was just one part within a larger analysis pipeline. Although the model was successful at defining tree boundaries, the goal was to identify evergreens where foliage remains green and functional year-round.

Bacon employed a series of techniques and ancillary data to further filter the data and identify areas with evergreens that potentially needed treatment. He used a variety of methods to aid in these selections, including the size of tree patches, the greenness of pixels within the imagery, infrared information within the source imagery, and

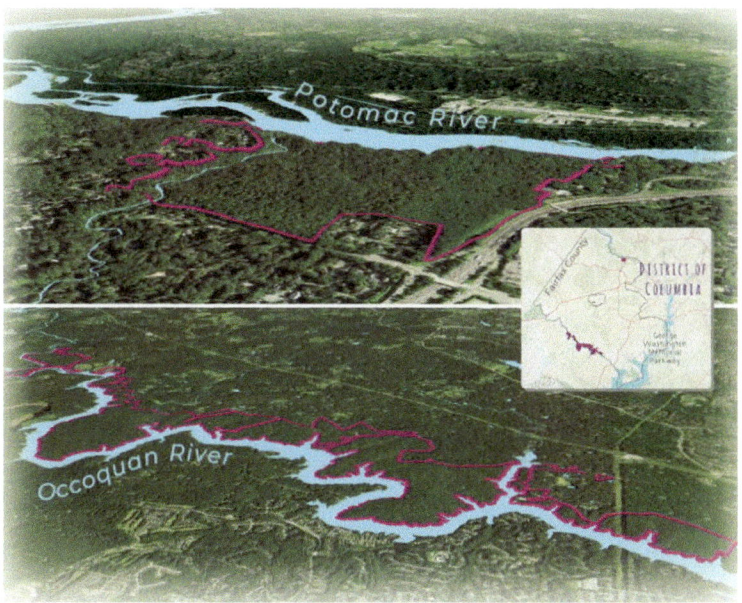

The study areas were adjacent to the Potomac and Occoquan Rivers.

lidar intensity values. This combination of factors allowed him to further distinguish between evergreen and deciduous tree species.

Working in a more precise target area

All the data processing results were queried to provide a comprehensive picture of not only where Fairfax County's evergreen trees were but also where hemlock trees were likely to be located. Once Bacon and the GIS Division made their selections in the imagery and generated outputs, the observations were converted to density surfaces. In the end, staff identified 33,110 evergreens out of 116,521 trees in the project zone—drastically narrowing the project to a more precise target area.

The GIS Division shared its results with the Forest Pest Branch management team through a file geodatabase, and the urban foresters published the data to ArcGIS Online. This helped inform a treatment plan. From there, the Forest Pest Branch's mobile team used ArcGIS Collector, which has since been replaced by ArcGIS Field Maps, to locate and treat infected trees in the study area.

While analysis of imagery data formed the core of the treatment plan, employing a well-rounded combination of GIS tools kept stakeholders informed and helped them quickly and efficiently allocate resources. Going forward, the GIS Division will use imagery and conduct additional on-the-ground work to monitor the effectiveness of the treatment.

Using one model to train another

The results of the hemlock woolly adelgid operation laid the groundwork for Fairfax County to do additional projects that involve using imagery to detect evergreen trees.

"Automating the process of mapping trees of interest and doing that successfully with help from ArcGIS software and apps established a benchmark for more efficient identification and monitoring moving forward," said Bacon.

An updated model mapped more than 36,000 evergreen trees on the Mason Neck peninsula. The 2022 natural-color imagery that was used to train the model is shown on the left, and 2021 infrared imagery to enhance the identification of evergreens is shown on the right.

In 2024, the county used a selection of evergreen tree boxes to train a new deep learning model to find only evergreens. Staff employed the same three-band, three-inch-resolution imagery from 2022 that they used in the hemlock project to train the new model by creating image chips containing a variety of evergreen samples.

The resultant model was then run against images of the Mason Neck peninsula in southern Fairfax County—an area of approximately 9,000 acres that is nearly three times the size of UFMD's hemlock project. The model mapped more than 36,000 evergreen trees throughout Mason Neck, performing best in underdeveloped areas that have dense clusters of evergreens.

Once again, using GeoAI to automate the process of using imagery to detect characteristics that separate evergreen trees from deciduous ones, UFMD was able to effectively plan more targeted mobile work to help in the fight against woolly adelgid on hemlock trees in Mason Neck.

This story originally appeared in the Winter 2025 issue of *ArcNews*.

Using GeoAI to Inventory ADA Curb Ramps Saves Significant Time and Money

Douglas County and the City of Omaha, Nebraska

A big part of unlocking the value of GIS technology is constant innovation, including learning new capabilities and searching for opportunities to apply them. Douglas County, Nebraska, has established an innovation culture in its GIS group—it constantly expands its knowledge and use cases of the technology across the organization. With about 585,000 residents, Douglas County is the most populous county in Nebraska and home to the city of Omaha.

Like all US government agencies, Douglas County and the City of Omaha must abide by the stipulations in the Americans with Disabilities Act (ADA) of 1990. This important legislation protects people with disabilities in many areas of public life. One of the accessibility standards associated with the act covers curb ramps, where sidewalks end at a curb and provide access to people crossing the street. These can be dangerous for people with disabilities, so there are ADA-specific design standards that must be followed in their construction. Over the years, as the Omaha Public Works Construction Division has been improving its roadway intersections and adding new intersections, each of these projects has included installation of ADA curb ramps.

As the city installed the ramps, it wanted to collect asset information and location to help with its asset life cycle management program, powered by Cityworks. Field data collection requires too many resources. As the city was already collecting aerial imagery every two years, it had administrative staff in the office enter the ramps and their attributes into its GIS using the aerial imagery as a reference. Over the years, this process lacked consistency and data standards, which led to poor data quality for the ADA curb ramp layer.

County GIS staff had become aware of GeoAI capabilities within ArcGIS, including deep learning models for extracting features from imagery, and had been looking for an opportunity to try them out. The ADA curb ramp inventory seemed like a great use case to test these capabilities, since ramps can be recognized by the human eye. Using ArcGIS Pro and the county's one-inch-resolution digital aerial imagery in a mosaic dataset, Steve Cacioppo, a senior GIS analyst at the county, set about applying a deep learning model to help solve this problem.

Developing a deep learning model takes trial and error, and Cacioppo worked through many scenarios using samples of data. One scenario narrowed the focus of the model to only areas where intersections and ramps are located, since there is no need to analyze

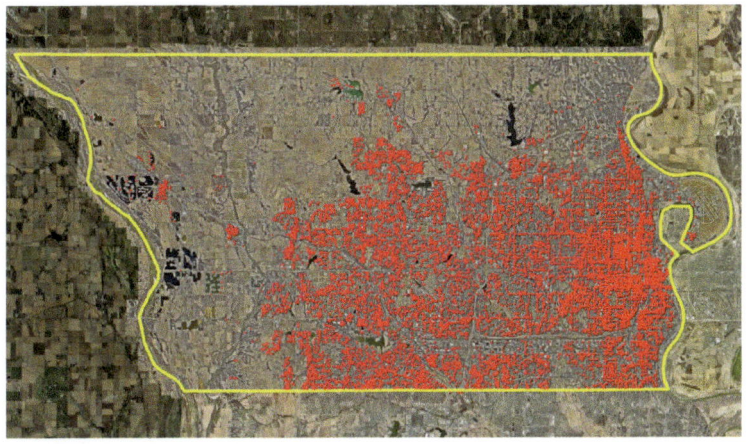

A countywide view of all ADA-compliant curb ramps that were located using the model are shown in red.

every square inch of the county. Multifamily housing and commercial areas were also isolated, since they have curb ramps. The county has a layer identifying areas without sidewalks; these areas were not analyzed. The model was not 100 percent accurate—some ramps were lost in shadows, and initially it identified some car sunroofs as curb ramps—but these issues were minor and correctable.

Once the model was finalized and run, it identified 34,183 ADA curb ramps. The original inventory included 16,775 ramps, so the improvement was substantial; the number of ramps was more than double the initial number, and the Construction Division was excited about this big improvement in data quality. The time savings have been considerable. On average, it had taken county staff one to two minutes to add a ramp into the GIS. The GeoAI model identified the ramps in about 12 days on a PC using ArcGIS Pro. With the curb ramp inventory completed, staff can now identify which remaining crosswalks need ADA curb ramps and can use GIS to prioritize their installation for inclusion in the Capital Improvement Program (CIP), as well as review those projects through an equity lens, such as the ArcGIS Solutions Social Equity Analysis configuration.

The resulting ramp data was shared with other municipalities in the county, so they know where their ADA curb ramps are located. Now that the county has this process in place, when new aerial imagery is flown every two years, county staff will be able to update the inventory in an automated fashion. Accurate management of these critical assets is important for the county in terms of its safe streets and Vision Zero programs, and the GIS team is already supporting that effort with ArcGIS HubSM and ArcGIS Dashboards.

This innovative experience has expanded the county's appreciation for GeoAI, and staff are looking for additional ways to apply it, including feature extraction of swimming pools for property appraisal and health department inspections and testing to see

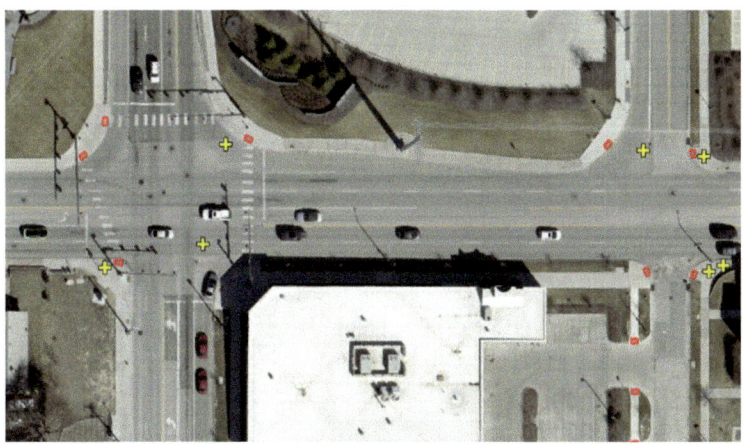

This image shows the GeoAI-extracted ADA curb ramp features, outlined in red, compared with the manually entered point data, symbolized by yellow crosses. It also shows how the GeoAI model missed two curb ramps because of shadows.

whether this is applicable for pavement markings, sidewalks, and bike lanes. The county also wants to apply GeoAI to its lidar data to extract trees and the edges of pavement. Other improvements being considered are moving the imagery to ArcGIS Image Server or ArcGIS Image for ArcGIS Online to improve GeoAI model and other geoprocessing performance.

The responsibilities of a GIS professional include constant learning and innovation, and Cacioppo and the rest of the Douglas County staff take that to heart. "This project really served two purposes for us. First, it was our initial attempt at creating a deep learning model from scratch. The success of this model will pave the way for future Esri deep learning models. Second, this model provided a time-saving approach to rapidly collecting ADA curb ramps for the City of Omaha. It will be used anytime we get new imagery to continually update the ADA curb ramp inventory," said Cacioppo.

"This model has been a great tool for identifying ADA curb ramps that we missed in our initial collection," agreed Todd Spark, City of Omaha maintenance superintendent. "This time-saving model allowed us to shift human labor from this project to other projects that needed more assistance."

The curb ramp project is one example of how Douglas County staff are being creative with GIS to be effective at the county level and in the community.

This story originally appeared as "County Innovates Using GeoAI to Inventory ADA Curb Ramps and Saving Significant Time and Money" at esri.com.

Improving Roadways Using Drone Imagery and Machine Learning

Utah Department of Transportation

The Utah Department of Transportation (UDOT) is responsible for planning, designing, building, maintaining, and operating the state highway system. It maintains major roads and moves traffic over long distances, including the interstate. Team members at UDOT believe that good roads cost less, and with proactive preservation, they can maximize the value of their infrastructure for today and the future.

Understanding the location and condition of assets helps UDOT make every dollar go further by prioritizing maintenance and construction efforts to provide the greatest good to the traveling public. By combining asset location data with other roadway characteristics, such as traffic volume, speed limit, and crash frequency and severity, UDOT can prioritize projects to maximize their utility. Ensuring that the digital representation of these assets is as accurate and up-to-date as possible is critical to their usefulness.

The challenge

UDOT uses mobile lidar to collect pavement information every year. Pavement striping and markings are important to track because they communicate information to road users about a variety of important driving tasks as no other traffic control device does. A recent Federal

Utah's wide-ranging climate makes inspection and maintenance tasks challenging.

Highway Administration report shows the US spends approximately $2 billion annually on pavement marking. Corey Unger, the spatial technologies manager at UDOT, said, "It's not only beneficial to human drivers but also to emerging autodriving technology."

While pavement marking is critical to maintain, Utah's wide-ranging climate makes this task difficult. The varying weather creates a tricky environment for planning and implementing the optimal type of striping, and UDOT has received criticism as a result.

The solution

To better maintain assets such as striping throughout their entire life cycle, UDOT is using digital delivery to transform the way projects are managed. This offers the organization new ways to understand, view, and use project design data in the field. Digital delivery for UDOT means the digitization of the project delivery process. Data is compiled and delivered digitally in each stage of the life cycle of a project, from its design to the construction to the return of the data to asset management for future project planning and execution.

This transformation led UDOT to create a digital twin, or a digital representation, of all the physical assets that make up Utah's transportation network. It's part of a scalable information management strategy.

"The purpose of digital delivery is to move UDOT away from static data and instead use dynamic data to represent the current state of every asset," Unger explained. Data is collected by drone using Site Scan for ArcGIS, which enables end-to-end drone management and direct integration with the ArcGIS system. It allows UDOT to easily overlay design data from the digital delivery process with imagery collected during and after construction on a project. This helps assess the accuracy of how a project was planned and implemented.

One example is UDOT's use of cut-and-fill analysis on the design surface compared with the elevation of the drone imagery collected for a port of entry project. This kind of analysis informs UDOT whether the road was built too high or too low, as either scenario can lead to dangerous conditions for vehicles, poor drainage, or other potentially hazardous situations. If the on-the-ground elevation is lower than what was designed, it's considered a cut, and if the on-the-ground elevation is higher than the design, it's considered a fill.

The same method is used to compare the designed pavement marking with where it was placed. This data gives the project inspector the ability to verify the placement of designed features in the construction as part of UDOT's digital twin effort. Once the accuracy of these placements is verified, the markings can be entered into the organization's asset management system, where UDOT can track the asset's life cycle and plan and schedule maintenance tasks accordingly.

Although data collection through Site Scan helped UDOT get closer to achieving a digital twin, the process still required it to

UDOT uses cut-and-fill analysis on drone imagery and the design to compare on-the-ground elevation with the design elevation.

manually inspect and pull assets from the imagery. "After pilots would fly sections of roadway, analysts would take the imagery and manually delineate pavement conditions and striping information in ArcGIS Pro," Unger said. "This was very time-consuming." So UDOT began exploring an automated process where it could send drone data collected from Site Scan directly through a GeoAI machine learning model and then extract striping data straight into its database.

UDOT reached out to Esri through the Esri Advantage Program to get its help creating a machine-learning model and methodology that would move the department from manual digitization of assets to asset detection using automated processes. Esri reviewed UDOT's program and provided a recommendation for best practices. Esri suggested rule-based modeling, which requires a predefined set of rules that can be applied to an image, so an exploratory assessment was performed.

During analysis of UDOT's drone data stored in Site Scan, team members examined two formats: orthorectified imagery and point

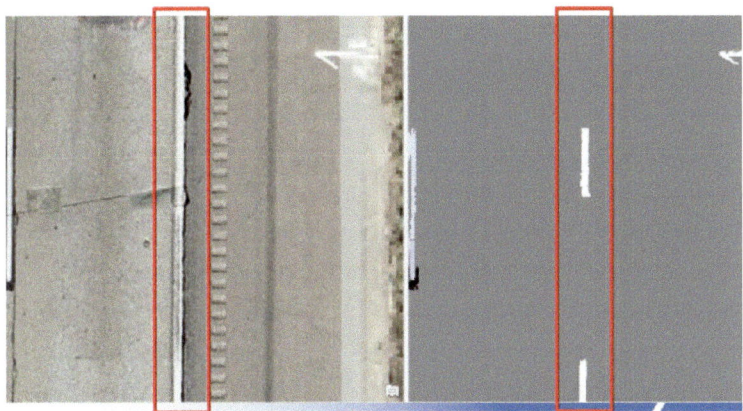

The drone image on the left is compared with the image on the right, which resulted from analysis. The white stripe that's outlined in red in the right image should be a solid line, like the shoulder marking that's outlined in red in the left image. However, the striping in the right image is broken, likely due to the darker color present in the stripe in the left image.

clouds. They decided to evaluate data in the corresponding workflows for various assets, including lane striping, rumble strips, pavement distress, signs, sound barriers, and traffic barriers.

In one example from the initial striping extraction results, a drone image of a road shows a solid striping line, while its coordinating analysis image shows broken striping lines. This can indicate poor striping conditions, but it requires further study. However, it's still valuable data for maintenance teams so they can target areas where repairs may be needed.

Another example from preliminary results shows a section of freeway where you can differentiate between concrete and asphalt sections, as well as define the striping. Again, this is important information for an accurate inventory of surface type and to help maintenance teams know what type of surface they're going to be repairing.

The result

Using remotely sensed data, coupled with AI and machine learning targets, enables quicker and more accurate data collection on infrastructure assets, so infrastructure owners can more quickly address maintenance needs. Site Scan and the Esri Advantage Program have been critical in digitizing asset inventory and developing machine learning models that will continue to save the organization time and money while also delivering a positive message to the community.

UDOT's goal is to have an automated method for extracting asset conditions from drone imagery to keep asset inventory as up-to-date as possible. The department plans to use these models as part of its efforts to refine its asset extraction process and expedite the completion of the digital twin. Deep learning, artificial intelligence, and drone imagery collected with Site Scan are important elements of UDOT's strategy for success.

This story originally appeared as "Utah Improves Roadways Using Drone Imagery and Machine Learning" at esri.com.

Drones, Maps, and AI Unite City Functions

City of Vilnius, Lithuania

Digital twins are an increasingly common tool for city governments to understand and manage city assets. The twins act as 3D aids for city planners, providing responsive models that show the likely effects of proposed changes.

Vilnius, the capital of Lithuania, has taken the model a step further. In addition to its utility as a planning tool, Vilnius's digital twin also provides a way to strengthen and streamline city services.

Snow removal, a perennial necessity in one of Europe's more blizzard-prone cities, is a key example. In January, the first major

The high-rise buildings in downtown Vilnius are surrounded by grassy fields and recreation opportunities. Photo courtesy of the City of Vilnius.

blizzard of the year battered Vilnius, but the city had most streets cleared within four hours. City officials used a combination of AI, location technology, and aerial imagery to accomplish this.

Moments after the storm ended, drone aircraft flew overhead, gathering images of the streets. An AI model trained to recognize snowy streets did a quick analysis. The result was displayed on a map of the city that formed part of a digital twin, built and maintained using a GIS.

Stasys Savilionis, head of the data management group at the Vilnius Data Center, the office that manages the digital twin, displayed a postblizzard map. Streets were color coded, based on the AI model's measurements.

"If a street is red, it hasn't been cleaned," Savilionis explained. "If it's yellow, that means it still needs to be inspected by a person. And if it's green, that means the AI identifies that it's been completely cleaned."

Creating a real-time digital twin

Vilnius complements its twin with drones to enable real-time analysis. The city's fleet of four drones gathers imagery and other data that can be projected onto maps and analyzed directly or using AI models.

Savilionis pulled up another map, this one with municipal trash receptacles highlighted. As with snowy streets, the AI had been trained to recognize full trash bins. Those in current need of emptying were highlighted in red.

Maps inside the digital twin serve more than one purpose. In addition to helping the city manage municipal functions, they provide a record that lets officials compare services over time. The public-facing versions also provide transparency for Vilnius residents.

Vilnius uses a detailed 3D city model as the foundation for decision-making and as a public gateway for users to learn more about the city.

Training an AI model to prioritize attention

The relative ease with which drone footage becomes operational intelligence belies the massive effort used to train the AI model. Thousands of photos are used to teach the algorithm what to look for, a process known as deep learning. The deep learning models then ingest the latest drone imagery to return answers.

Once trained, the AI can spot priorities for service. These areas appear on dashboards for city leadership and guide the work of relevant city departments. For example, the city can assess road conditions by using the AI to highlight cracks, potholes, and other damage that may require attention.

The model can also recognize where traffic is backed up, where cars are parked illegally, and where large snow loads have accumulated on roofs. "That's especially useful for buildings that might be hard to access by foot," Savilionis said.

Savilionis's team tunes the AI model to consider different parameters in its priority ranking. Recently, they asked it to rank streets

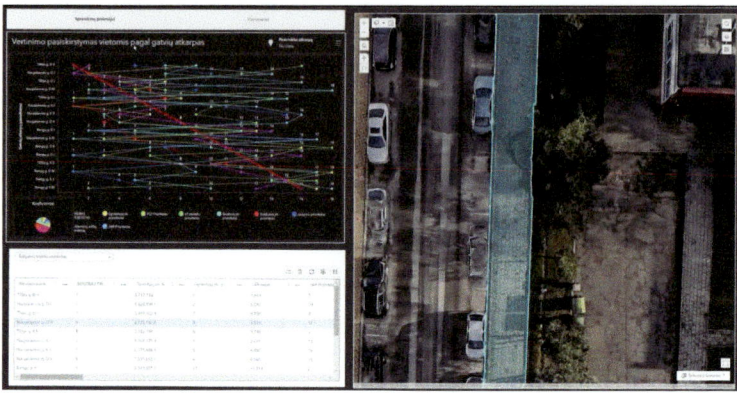

Drone imagery provides the input to assess condition of the pavement to prioritize repairs. Screenshot courtesy of the City of Vilnius.

based on their importance for pedestrian traffic. Before that, they tuned it to account for routes of cyclists and commuters. The AI model notes, for example, how close a street is to schools and hospitals and other places where many people need to travel.

"A street with really bad conditions might rank low on the metric because of its distance from health-care or educational institutions," Savilionis said.

He noted one street that had a high score for urgency because of its proximity to a playground. This information helps expedite repairs and can be shared with the public, increasing transparency and accountability.

Enhancing a shared awareness of the city

In addition to real-time analysis, Vilnius's digital twin helps officials understand how city operations might evolve over time.

With anonymized cell phone location data from mobile operators, the Vilnius municipal company responsible for public transit can map and visualize where people congregate and how they move at different times. This insight helps assess current transit needs and predict how mobility demands will change in the future.

"We get this data every day," Savilionis said, clicking between two versions of the same map. "You can see how a street in August looks very different from one in September when school is in session."

Like many cities, Vilnius is trying to move away from dependence on cars. Having a comprehensive view of all relevant data helps planners improve pedestrian routes and bicycle lanes and take measures to improve mass transit.

Savilionis displayed a map of Vilnius centered on a particular bus route. The map is composed of colored tiles, each representing the number of people in that location who depend on the route for commuting. The breakdown gives city leaders a clearer idea of whether more buses should be added to existing routes—or if entirely new routes are required.

"We are monitoring services provided to citizens," Savilionis said. "Our actual goal is to identify problems before our citizens do."

This story by Brooks Patrick originally appeared as "In Lithuania's Capital, Drones, Maps, and AI Unite City Functions" in the *ArcGIS Blog* on August 27, 2024.

As Sea Level Rise Threatens, Safeguarding a Sense of Place with a Digital Twin

Government of Tuvalu

Rising seas are slowly swallowing the Pacific island nation of Tuvalu—and the country's leaders are taking action. They are using a digital twin to understand climate impacts and preserve their culture.

Nine islands make up Tuvalu, home to more than 10,000 people and situated halfway between Australia and Hawaii. The country—on the forefront of climate action—changed its constitution to declare it would always be a nation, even if its islands are no longer inhabitable.

"We are looking at setting up the world's first digital nation because we are facing the real risk of our land disappearing," said Tuvalu's foreign minister, Simon Kofe, standing knee-deep in water while addressing the United Nations Convention on Climate Change (COP26) in 2021.

Digital twins come to life with a combination of imagery and GIS technology. With the digital twin, decision-makers can analyze details about a place, its people, and the environment. The 3D model serves as both a record and a platform to navigate incidents and changes. It also provides a means to experience the place within virtual reality headsets.

Kofe's speech, and the country's commitment to digitize itself in case it disappears, puts the world on alert to the vulnerability of island nations. Many others, including Grenada, have adopted similar reality-capture techniques to face a changing environment and record what makes them unique.

Capturing the whole of a vulnerable nation

Tuvalu is experiencing many climate hardships, including areas perpetually underwater, saltwater intrusion into drinking water, and coral bleaching. Because the depth of the ocean drops off deeply around the islands, the shores aren't susceptible to huge waves. But the islands have experienced cyclones, which erode shorelines and destroy homes. With the digital twin, Tuvalu can assess the dimensions of these ongoing challenges.

The first step is to collect location-specific data and regularly update it so changes can be measured. In September 2023, in concurrence with the United Nations General Assembly meetings in New

The drone team captured imagery from many places on the island, including at the end of the airport runway. Image courtesy of PLACE.

York, and with the support of the Secretariat of the Pacific Community (SPC) Digital Earth Pacific (DEP) program, the nonprofit PLACE signed an agreement with the prime minister of Tuvalu to collect detailed data for the total land area of 26 square kilometers.

"The Digital Earth Pacific program enables decision-making for the Pacific Peoples through the use of earth observations. PLACE data enhances the program by providing highly detailed and accurate images that support Pacific nations in addressing climate change, food security, and disasters," said Stuart Minchin, director-general of SPC.

In April 2024, the PLACE team traveled to the remote nation and collected images of Funafuti, the country's capital. They used a drone to capture aerial images and a spherical camera mounted to a truck to capture street-level views. This sensor pod camera with five lenses on the sides and top can also be attached to a backpack for capturing areas that can be reached only on foot.

PLACE's imagery data is processed to create 2D and 3D data products using ArcGIS Reality, a suite of photogrammetry software products designed to enable reality-capture workflows for sites, cities, and countries. The GIS tools match large groups of images taken by cameras along the many flights and routes. This set of imagery data works together inside GIS to create a precise 3D model.

The resulting immersive basemap forms the foundation of a digital twin. From there, the team can add an inventory of data about physical assets, trees, houses, and infrastructure. Within the digital twin, they use the AI technique called machine learning to automatically extract details from the images, such as building outlines and changes to the shoreline.

"The most exciting thing for us is what's possible with machine learning and desktop computers now," said Frank Pichel, partner and global field operations lead at PLACE. "The conversations we were having a year ago versus now are a world apart."

Part 2: Public Sector Applications   37

The imagery was analyzed to detect such changes as new development, as in this image comparison from 2019 on the left to 2024 on the right on the northern end of Fongafale. Image courtesy of PLACE.

With the digital twin, a flood model can be automatically created and overlaid with a population map. This awareness helps leaders assess changing conditions, engage residents in the crisis, and empower everyone to make informed decisions about their future.

Heightening pressure in low-lying areas

On every full moon when the highest king tide hits, the tarmac on the runway of Tuvalu's only airport floats away from its coral base. Water bubbles up through the porous island terrain, pushing it above the surface. "They have to drive bulldozers down the runway to get it back on the ground before any plane can land," Pichel said.

This regular occurrence illustrates Tuvalu's precarious situation—at the tipping point of rising seas. A global study that mapped land elevation in relation to coastal flooding estimates that, worldwide, more than 410 million people will have to leave their homes by 2100.

For its future, Tuvalu recently signed a treaty with Australia that will allow 280 Tuvaluans to migrate each year on permanent resident visas. "It's very sad to see that realization that everyone feels the need to have a foot somewhere else," Pichel said.

With the digital twin, future generations of the Tuvalu nation will be able to experience their homeland even if it sinks. While this move ensures a type of perpetual homecoming, it can also be effective in the present.

Digital twins are already being used to address the unique social, economic, and environmental vulnerabilities across an increasing number of small island developing states.

As virtual reality technologies improve, people may be able to experience Tuvalu as it was—to delight in its beaches and biodiversity-rich reefs. They may choose to visit the digital twin regularly, not just to reflect but to center themselves on their ancestral heritage.

This story by Dawn Wright originally appeared as "Threatened by Sea Level Rise, Tuvalu Safeguards Its Sense of Place with a Digital Twin" in the *ArcGIS Blog* on October 29, 2024.

Deep Learning Helps Automate Map Updates to Better Serve Citizens

Government of Kuwait

Kuwait Finder, a mobile location app built by the Kuwaiti government, was a success when it was first released in 2013. Among its many achievements, it was a triumph of data-gathering and data-processing, offering authoritative turn-by-turn directions for Kuwait, a locale whose system of address numbers and street names can be confusing.

By 2020, Kuwait Finder had amassed 750,000 users. Within the country, that level of saturation makes Kuwait Finder a more popular wayfinding tool than Google Maps.

This popularity presented its own challenges, including how to keep the application and underlying data up-to-date. Per capita, Kuwait is the fourth-wealthiest country in the world, and with the Kuwait National Development Plan goal to increase infrastructure expenditures by 11 percent, change is already under way. Construction projects—including the world's longest causeway, a new airport passenger terminal, and a new 500,000-person residential area called Silk City—amount to $500 billion in active investments, including major changes to transportation networks and other infrastructure.

To remain trusted and reliable, Kuwait Finder must capture and reflect this dynamism, with an up-to-the-moment authoritative geospatial rendition of the entire country.

Turning to automation

Initially, to create Kuwait Finder in 2012, a five-person staff at Kuwait's Public Authority for Civil Information (PACI) tapped into the country's long-term investment in GIS technology. The GIS team pulled together data from various ministries, as well as PACI's internal paper maps and AutoCAD files, to capture the current basemap. With the objective of updating the construction of new buildings and changes to the city's streets, they studied satellite imagery to extract details for the map. But by the time the crew found the changes and input them into the Kuwait Finder database, the infrastructure of Kuwait City had changed once again.

"Where are the new streets? Where are the building footprints?" Maher Abdel Karim, PACI's GIS consultant said recently, summarizing the challenges. "Once we were finished updating them, there would be new satellite imagery we'd missed, and we'd have to repeat the process again."

PACI needed to make necessary changes to Kuwait Finder in a way that was quick, inexpensive, accurate, and simple—the more

Few old buildings still stand in Kuwait City, where construction cranes are a common sight.

automated, the better. Compared with the vast resources available to a company such as Google, PACI's size and budget were limited.

PACI researched the viability of artificial intelligence to provide a solution.

"We thought, how can we use AI to automate the process?" Abdel Karim said. "We wanted to use machine learning to extract street data and building footprints from the satellite imagery while using the minimum amount of human input."

Deep learning to the rescue

Deep learning, a powerful form of AI, involves teaching a computer to detect patterns in large amounts of data and to recognize and extract just the information you want. If done right, the algorithm acts quickly and thoroughly and even finds changes that human intuition would miss.

PACI's GIS team needed to teach the computer how to recognize building footprints from satellite data and note which ones were new since the last batch of satellite images. Elsewhere in the Middle East, an oil-and-gas company was using machine learning to alert it if any new structures were being built near its thousands of miles of

Building footprints digitized manually on the left compared with the deep learning algorithm outputs on the right.

pipeline. PACI's task was a bit more complex as it wanted the computer to inform it about anything new across the entire country.

If properly trained, a machine learning algorithm can move conceptual mountains. At the beginning of the process, however, the computer needs to be taught to "read." This requires the same patience and skill required to teach a young child to recognize letters, then words and sentences, and finally complex thoughts.

PACI needed to establish a ground truth, a common geospatial framework that would encompass the existing database. After much trial and error, the staff was able to train the model using 75 square kilometers of data to provide input for it to scan 3,000 square kilometers of satellite imagery.

Knowing the local vocabulary pays off

For PACI, the main challenge was giving the program enough information to recognize buildings and streets.

"The contrast between land, streets, and buildings was very minimal," Abdel Karim explained. "It's hard even for a person to differentiate among them by eye. And our threshold for saying that our model worked was very high."

Often, with shadows cast by the sweltering summer sun, building footprints did not easily align with the images captured by the satellite. The darkened overlapping edges frequently confused the program.

Streets posed their own unique problem. They had to be recognizable individually but also as part of a coherent grid. PACI's staff had to ensure that all streets in the imagery were connected by common center lines—and that the lines in the new images fit seamlessly with those in the existing street network.

"You really have to take your local experience and use it as a 'flavor' on top of the satellite imagery," Abdel Karim said. "That's what makes it possible for machine learning to perform these calculations."

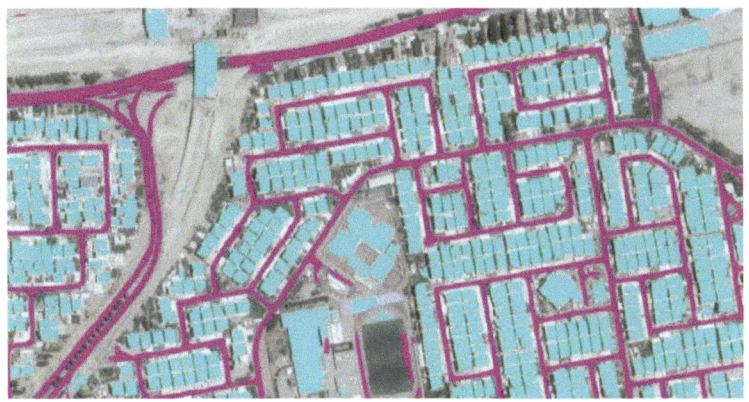

The algorithm for aerial imagery recognizes both roads and buildings.

The teaching process required serious work. But once the machine learning model was ready, the rest was simple. PACI devoted time to training and fine-tuning the model, knowing that this work would pay dividends in the future.

The model and training dataset can be used today to update the database in about three hours and provide even more accurate maps than before. And this model will live on and can continue to be modified and enhanced by PACI, taking into consideration new architecture styles, building types, or other new features.

Reaching the payoff

The model could now do, before lunchtime, a task that previously took five humans a year to complete. The time spent to train the machine to extract features from satellite imagery to its GIS repository had paid off.

"You need to have a lot of patience," Abdel Karim said. "It can be a fairly long, iterative process to train, predict, evaluate, and then go in and do it all over again. It's a knowledge investment. You're investing in your staff, who are working, studying, and learning how to do this, to get the desired output."

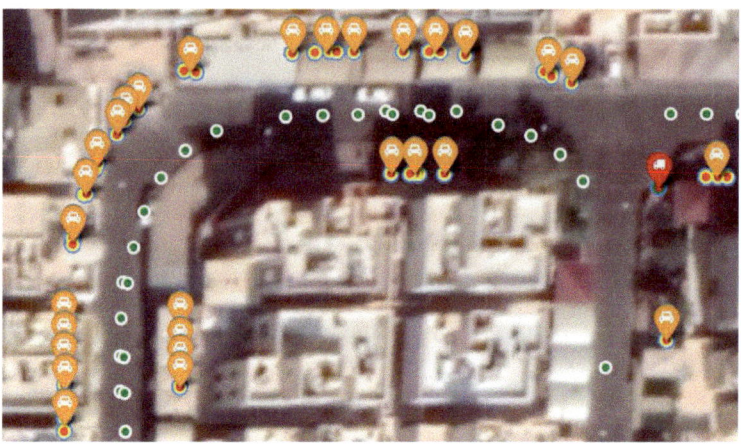

The output from the street-level imagery can distinguish between parked cars and trucks.

The return is worth the investment. The organization now has the confidence that it can keep the Kuwait basemap and Kuwait Finder updated in a timely manner so that it can continue to be the authoritative source for location information in Kuwait. The time and cost savings from this automation will allow the small team at PACI to keep innovating to stay ahead of the competition. The lessons learned from using deep learning and remotely sensed data will also feed new ideas for PACI staff and foster more innovation.

"We will continue to use machine learning to create new features and new layers for our GIS repository that people really need and are looking for," Abdel Karim said.

Turning AI eyes to the streets

Similar to Google Street View, PACI invested in its own mobile mapping vehicle to capture street-level images to help Kuwaitis in their wayfinding. Seeing an address in the same context as looking through a car window helps users pinpoint the destination and better visualize the location.

"Now, the idea came to us to use machine learning to extract features from this street-level imagery," said Maher Abdel Karim, PACI's GIS consultant. "To add things missing in our database, such as traffic lights and signs."

PACI also envisions using the street-level imagery to help monitor the progress of current projects, as well as to guide the city's maintenance operations.

"We could find the streets that have bad conditions," Abdel Karim said. "Instead of doing mass street rehabilitation, we could target only the part that has a problem."

PACI is now adapting the same machine learning training tools it used for satellite imagery to this new street-level task.

"We are trying to build this kind of machine learning–supported data approach so that we can feed our models with data and specify the output that we are looking for," Abdel Karim said.

This story by Linda Peters originally appeared as "Deep Learning Helps Kuwait Automate Map Updates to Better Serve Citizens" in the *ArcGIS Blog* on May 12, 2020.

GIS and AI for Precise Damage Assessments

Esri

When wildfires ravaged Lahaina, Hawaii, the scars left behind weren't just physical but emotional and societal. The subsequent challenges of conducting a damage assessment helped underscore a pressing need for improved tools and methodologies. Although traditional methods have their merits, the scale and severity of such disasters demand something more efficient. So, Esri developed a new damage assessment deep learning model to meet the need.

Historically, in the wake of a disaster, conducting a damage assessment has been a time-consuming, manual endeavor. Damage assessment teams need to cover vast areas, often under hazardous conditions, to document and assess damage. This not only prolongs any potential response times but also can sometimes result in inconsistencies due to the sheer scale of the disaster.

Esri's deep learning model was designed with the primary goal of addressing these challenges. By processing high-resolution satellite and aerial imagery, the new damage assessment model can identify patterns of destruction, differentiating between damaged and undamaged structures. This automation speeds up the assessment process with surprising accuracy. Here, we will provide a high-level overview of the workflow and explain where you can locate the model to start using it yourself.

Practical application in Lahaina

In the wake of the Lahaina wildfires, we developed and put to the test a new damage assessment model. What would traditionally take hours of labor and manual surveying was achieved in a fraction of the time with 95 percent accuracy. Rapid damage assessments with this level of accuracy revolutionize postdisaster assessment processes by providing access to previously unobtainable decision-support information during an initial response. For instance, with rapid and accurate damage assessments, authorities can prioritize areas for search and rescue and immediately allocate resources to heavily affected areas.

To classify damage, you must have building footprint information for the area of interest. This information allows the damage assessment model to classify buildings as damaged or undamaged by creating boundaries for referencing and labeling.

The accuracy of the building footprint layer is also important and may vary depending on the method used to create the data. The

The Deep Learning Model to Extract Building Footprints.

building footprints should be accurate to the incident imagery you are planning on using to run the model. If building footprints don't exist, the user needs predisaster imagery so that they can use different AI models to extract building features. To provide additional context for the buildings, the building footprint data Esri used also indicated building occupancy type. This allows the model to help quantify the number of homes and other infrastructure that may be damaged.

For help in extracting building footprints from imagery, Esri offers a high-resolution model in ArcGIS Living Atlas: the Deep Learning Model to Extract Building Footprints.

After obtaining building footprint information, the next step is to locate postdisaster imagery for the affected area. Postdisaster imagery can come from a variety of sources, including satellites or uncrewed aerial system technology. Obtaining postdisaster imagery from a satellite sounds difficult or prohibitively expensive. Neither is true. Resources are available to help you learn about requesting satellites to capture your area of interest and find out how easy it can be.

Once postdisaster imagery is obtained, it is best to refine the deep learning model. For Lahaina, we manually inspected 500 buildings from the postdisaster imagery, classifying them as either damaged or undamaged. This helped refine the model to different geographies, account for differences in building size and shape, and incorporate different extents of damage into a model's output. In short, we took our postdisaster imagery and conducted a drive-by or windshield survey to help the model classify buildings.

Once the model is refined, users can use ArcGIS Pro or ArcGIS Online to run the deep learning model and produce their own damage assessments. Esri published the results of the analysis and the deep learning model in ArcGIS Online in hopes of spurring efficient collaboration, creating new tools for incident response, and making GeoAI available on a wider scale.

Part 2: Public Sector Applications   49

This image shows additional training samples the team created to help refine the preexisting damage classification model. Red building footprints signify damaged property, and blue are undamaged.

The model's results can be placed into a web map, dashboard, or web app, depending on the user's need. In the case of the Lahaina fire, a dashboard was chosen so that the results could be visualized, and organizations could use the dashboard to help guide response and recovery efforts as well as inform public outreach.

Lahaina Damage Assessment Dashboard.

The way forward

The tragic wildfires in Lahaina served as a stark reminder of the vulnerabilities we face and the importance of timely and accurate post-disaster assessments. Like most AI applications in wildland fire, deep learning is not a replacement for on-the-ground verification and expertise but rather a tool that augments human capabilities.

As wildland fire and disaster management continue to evolve, the blending of these advanced models with traditional methods will serve to refine our disaster response mechanisms. As technology continues to progress, our goal remains steadfast: to support communities in their time of need, ensuring that they have the best tools at their disposal for disaster preparedness, mitigation, response, and recovery.

This story by Anthony Schultz and Jarell Perez originally appeared in the *Esri Industry Blog* on November 30, 2023.

Getting the Most of GeoAI in Emergency Management
Esri

AI is often misunderstood in emergency management. On one end, practitioners are eager and ready to embrace AI's capabilities that can help increase productivity. On the other end, uncertainty about AI may cause anxiety, delaying adoption of the technology.

To ensure that AI's capabilities contribute positively to our work, we can establish boundaries and parameters to integrate into well-architected information technology (IT) systems. When adopting AI, its outputs become more reliable by enacting and incorporating standard IT practices (such as security, performance, scalability, and reliability), thus making the technology a trusted source to aid decision-making. Although incorporating tried-and-true IT security parameters may help alleviate concerns, understanding AI's capabilities and limitations remains the most significant barrier to full industry adoption.

Many emergency management organizations use AI's text-generating capabilities—such as large language models (LLMs) accessed via chatbots—for planning and preparedness tasks. However, extending AI's use to more advanced applications currently feels more like the industry's "final frontier." Market research unveils one likely reason—there is an absence of accessible and easy-to-understand information on how AI could be safely applied throughout emergency management programs. Furthermore, we are often misled into thinking that AI is "new" when, in fact, emergency managers have had access to AI-enabled systems since at least 2008.

It is true that technology is advancing rapidly, yet with the right constraints, training, and application, AI can be used safely and effectively.

Esri has been working on integrating AI throughout ArcGIS for years. This story aims to share information and knowledge in an easy-to-understand manner on ways emergency managers can use ArcGIS—an AI-enabled system—for pre- and postdisaster workflows.

ArcGIS—an AI-enabled system designed for modern enterprise

In emergency management, AI can support preparedness, mitigation, response, and recovery. Although generative AI models have grown in popularity, primarily through using LLMs to create new data and text, such as natural language, emergency managers also have access to machine learning and deep learning models for tasks including predictive modeling, object tracking, and imagery analysis.

ArcGIS is a foundational enterprise technology used by both business and government organizations. As a GIS, it provides location intelligence, contributes to the digital transformation of an organization, integrates with the Internet of Things (IoT), helps create digital twins, and enables teams and organizations to collaborate, make decisions, and act. Emergency managers have been using this system for decades to map and understand the world around them to aid in response and recovery efforts.

However, what may surprise some emergency managers is that AI in ArcGIS started with the inclusion of machine learning tools in 2008. In recent years, ArcGIS has been integrating smart assistants, the Text Segment Anything Model, vision language foundational models, and, most recently, new generative AI assistants.

In nearly all these cases, AI was architected directly into ArcGIS without the need for any additional configuration.

This means that emergency managers who have used GIS for

data analysis, imagery analysis, and prediction modeling have been successfully integrating AI into their programs for nearly a decade.

Let's break down some scenarios where emergency managers could take advantage of these capabilities within their programs.

Applying GeoAI and AI assistants into emergency management workflows

Predisaster

One of the more common ways to use AI predisaster is for predictive modeling and simulations. In emergency management terms, using ArcGIS to simulate coastal, riverine, or flashflooding uses GeoAI capabilities to conduct predictive analysis. Or when organizations work to identify structures in a floodplain that should be considered for mitigation projects, the building footprints visualized in map layers can use GeoAI's extraction capabilities. These extraction capabilities are pretrained models to find and identify objects that look similar to one another.

For emergency managers interested in modeling how the climate may change in the future, check out the National Oceanic and Atmospheric Administration (NOAA) Sea Level Rise Viewer, which is based on ArcGIS.

GeoAI can also assist emergency managers with situational awareness. For organizations with watch centers or watch desks, GeoAI pretrained models can be used to track moving vehicles or humans visible by CCTV. This capability is often used in security operations centers, where vehicles and VIP protective detail personnel are tracked and their movements visualized on a screen.

Postdisaster

Imagery analysis for damage assessments is a GeoAI capability that emergency managers use to expedite the recovery process. This capability was successfully applied in the aftermath of Hurricane

Ian, where geospatial damage assessments were conducted using AI, crowdsourcing, and high-resolution imagery from satellite, air, and ground. This process resulted in $78.3 million in assistance to survivors without requiring an in-person inspection.

Often, satellite, drone, or aircraft lidar images are collected after an emergency, showing aerial photographs of the extent and the scope of damage. Through AI's feature extraction and segmentation processes, viewers can classify and identify an object, such as a building or structure, and the machine can be trained to identify similar structures throughout the image. These pretrained models can then identify trends and patterns in the images of the disaster area and provide estimates of structure damage, extent, and debris amounts. You can try out the free tutorial on Classifying Objects Using Deep Learning in ArcGIS Pro.

Using pretrained models for object classification in ArcGIS Pro, GeoAI can detect patterns from imagery. These capabilities are frequently seen and used as change detection in images, such as "swipe maps," which show damage pre- and postdisaster. The models can also be trained to quantify and classify the extent of building damage (minor, major, destroyed, and so on), estimate debris volumes, and blur objects that may be sensitive.

For a more interactive experience with postdisaster imagery, generative AI can be used through a chatbot in ArcGIS Pro. In this example, users can prompt a map model with a question or statement in the prompt box, and the model will describe the image and trends for analysis in natural language. These capabilities are available through the Vision Language Context-Based Classification deep learning package that bridges ArcGIS Pro and OpenAI's vision language classification models.

Lastly, ArcGIS Survey123 can be configured to extract information from a photo and used to populate data in a form, using an AI assistant and computer vision AI. AI assistants can be prompted to help generate a Survey123 form, postdisaster. Auto Translate also helps translate survey components into several languages (see the section below on text translation for more details).

Search and rescue

Search and rescue missions can be streamlined using many of the same AI capabilities as damage assessments. As consumers of upstream data—such as a building footprint data layer and aerial imagery collected pre- and postdisaster—search and rescue teams can quickly identify and prioritize search areas using deep learning pretrained models. These models search for trends and indicators in images where damage is most extreme and suggest a prioritized list of geographic areas that should be searched first as an output.

As search and rescue teams are deployed, mobile operators can then use the mobile capabilities of Survey123, ArcGIS QuickCapture, and ArcGIS Field Maps to feed real-time data from the field to an operations center, capturing geolocation, text, images, and video. As this information is fed back into a common operating picture, search and rescue teams can swiftly and methodically ensure all affected areas are searched accordingly.

Text translation

Emergency managers frequently need to translate information into a variety of languages. Preparedness materials, such as educational pamphlets, brochures, and videos explaining family and individual preparedness techniques, are often translated. Emergency alerts, messages, emails, and broadcasts are also translated to reach as many community members as possible.

Recently, the Federal Communications Commission adopted requirements that participating commercial mobile service providers support templated Wireless Emergency Alerts (WEA) messages in 13 of the most common languages in the US.

One of Survey123's latest features within ArcGIS is Auto Translate, which allows respondents to easily translate question labels, choices, hints, and all text into the language of their choice using machine translation. The Auto Translate feature has been implemented into ArcGIS following strict privacy and security guidelines, ensuring that no data is ever shared with third parties, complying with ArcGIS Online privacy and security requirements.

An end-to-end system

ArcGIS is an end-to-end AI-enabled system designed and architected to enhance productivity and accelerate decision-making. It provides capabilities for creating, managing, analyzing, mapping, and sharing all types of data and delivers a decision-making edge through a geographic approach.

As an AI-enabled system, ArcGIS is designed for modern emergency management enterprises. It provides tools for automated data creation, extraction, and processing; advanced spatial analysis and visualization; predictive analytics and forecasting; and real-time data integration and processing. AI in ArcGIS is charting a new course for the future of GIS, helping to reshape the world in unprecedented ways.

Do you want to get started using GeoAI or AI assistants? Here are two things emergency managers can do today: (1) Access ArcGIS Living Atlas to begin using pretrained models in ArcGIS (seeing these models in action may help emergency managers become familiar with how GeoAI works and how its outputs can help inform their programs) and (2) challenge their GIS professionals to explore the learning packages within ArcGIS Pro and begin incorporating GeoAI for all data, imagery, and modeling tasks.

A version of this story by Carrie Speranza originally appeared in the *Esri Industry Blog* on February 11, 2025.

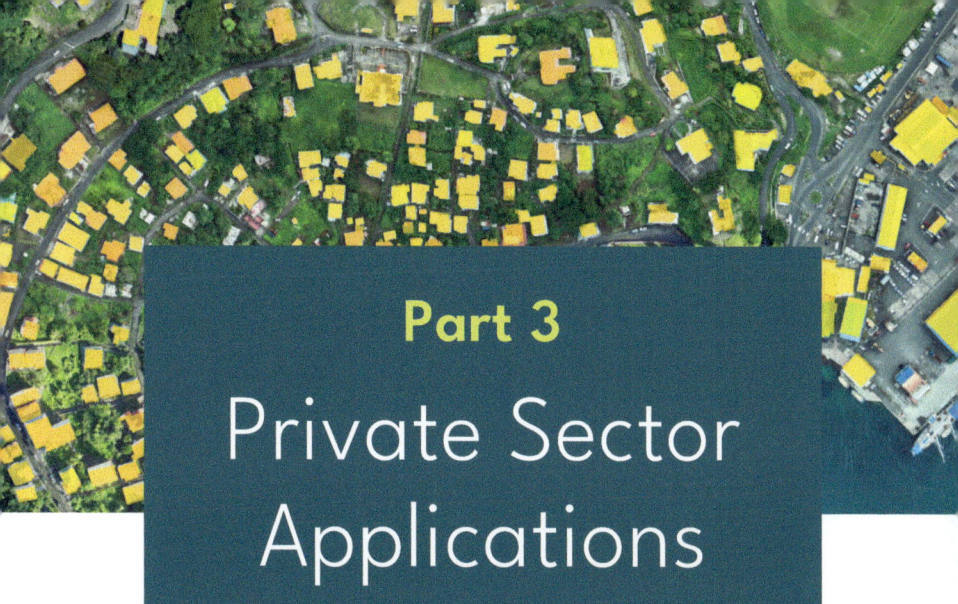

Part 3
Private Sector Applications

In the private sector, GIS experts and business leaders can use GeoAI to integrate spatial intelligence into strategic decisions. For example, energy companies can collaborate with their GIS teams to help answer complex questions while analyzing environmental conditions, regulatory requirements, grid infrastructure, and other key factors. GIS specialists can also help real estate, logistics, and retail companies transform their spatial analysis using GeoAI. Across the private sector, experts can guide GeoAI to help process complex location data, evaluate multiple scenarios, and recommend optimal solutions more quickly.

GIS in action

This section presents real-life stories about how GeoAI can help private-sector organizations increase efficiency, improve risk management, and more.

Deep Learning Model Unlocks Potential of Solar Energy Development

Pivot Energy

Solar power presents an immense opportunity for generating sustainable and green energy. However, realizing its full potential requires identifying suitable locations for solar panel installations.

Pivot Energy, a national renewable energy provider with its headquarters in Colorado, needed assistance locating parking lots across various areas of interest that would be appropriate for potential solar panel installation. Using advanced GIS technology, a team from Esri partner Platte River Analytics helped Pivot Energy do this accurately and efficiently.

By using a deep learning model from Esri, the team at Platte River Analytics extracted parking lot surfaces from high-resolution imagery. The team then used geoprocessing tools in ArcGIS Pro to conduct more precise measurements and calculations of potential sites.

The results of this analysis provided Pivot Energy with invaluable information, empowering staff to make data-driven decisions and plan solar power adoption efforts more effectively.

The benefits of parking lot–based solar development

Parking lots offer significant yet underused space for solar power development. They have key characteristics that make them ideal

Using Esri's Parking Lots Classification – USA deep learning model in ArcGIS Pro, the team conducted a test study of parking lot extraction in Golden, Colorado.

for generating solar energy, such as large surface areas, unobstructed exposure to sunlight, and proximity to electrical infrastructure connections. In addition, paved parking lots typically have very low slopes, are designed to drain, don't compete with other land uses, and aren't in full use all the time.

Solar-powered parking lots can provide numerous environmental and economic advantages to lot owners and communities as well. They can be quickly equipped with electric vehicle (EV) charging stations. This not only enables EVs to be powered directly by solar-generated energy, promoting clean transportation, but it also reduces electricity costs for lot owners and operators and opens avenues for potential revenue generation through energy sales. In addition, offering EV charging to customers attracts car owners who will spend money in the area while charging their vehicles.

All these are reasons why Pivot Energy saw the need to scour parking lots across the East Coast and, eventually, the nation to see which ones might be ideal for solar energy development.

A quick, automated way to detect suitable parking lots

To automate the detection of parking lots in Pivot Energy's areas of interest, the team at Platte River Analytics relied on Esri's Parking Lots Classification – USA deep learning model, available in ArcGIS Living Atlas of the World.

Developed by the Esri analytics team, this prebuilt model is trained to identify parking lots within sourced areal imagery. Like the more than 65 other deep learning models that the Esri analytics team has developed to detect objects ranging from Arctic seals to power lines, the Parking Lots Classification – USA model automatically extracts the assets from imagery without users having to invest time or money in training data or personnel.

For this project, the team at Platte River Analytics needed to use high-quality submeter data that allowed the model to identify and analyze land features as detailed as parking lots. The team acquired one-meter resolution National Agriculture Imagery Program (NAIP) imagery from the US Geological Survey's EarthExplorer web app. The imagery in this app has been acquired by the US Department of Agriculture during agricultural growing seasons since 2003.

After downloading the NAIP imagery, the Platte River Analytics team seamlessly integrated the deep learning model into its ArcGIS Pro workflow. The team processed the imagery with the model, which automatically identified parking lots across dozens of Pivot Energy's areas of interest.

The model was easy to use. The initial area of interest that the team investigated was the size of a large US city, and it took less than 12 hours to both download the imagery and process it in ArcGIS Pro.

To further analyze the identified parking lots, the team at Platte River Analytics used the Raster to Polygon geoprocessing tool in ArcGIS Pro to convert the raster outputs into polygons. This enabled the team to get more precise measurements and calculate the size of each lot, providing valuable information to Pivot Energy so staff could begin conducting feasibility assessments and get started with project planning.

From there, the GIS team at Pivot Energy was able to study regulatory factors—such as floodplains, tree cover, wetlands, and wildlife migration routes—around parking lots that were initially deemed acceptable for solar development.

Saving dozens of hours of manual work per week

Taking a machine learning–based approach to finding suitable parking lots for solar panel installation enabled staff at Pivot Energy to quickly make informed decisions regarding which areas and specific parking lots could work for this endeavor. By using advanced GIS to assess parking lot size and location, the developer can optimize project planning, ensure maximum energy generation capacity, and speed up the installation of solar infrastructure.

According to Rachel Mead, GIS manager at Pivot Energy, the process that the team at Platte River Analytics used to extract parking lots from imagery saved her own team more than 20 hours per week of manually searching aerial imagery for—and digitizing parking lots throughout—the company's areas of interest, which extend across the United States.

"It has been a huge time-saver having access to the deep learning models provided by Esri," she said. "By automating this ... we can save dozens of hours per week and realign that time to other projects."

This story by Andy Bohnhoff originally appeared in the Fall 2023 issue of ArcNews.

CEOs May Be Underusing this AI Capability

Bouwinvest

As business leaders roll out AI adoption strategies, they're undervaluing one key area, according to a new survey by *Fortune* and Deloitte. While more than 75 percent of surveyed CEOs expect AI to automate manual operations and increase efficiencies, just 31 percent expect it to improve risk management. But the right data and AI capabilities can be invaluable for assessing risk quickly, precisely—even years in advance.

Imminent physical risks, such as cracks in airport runways, can be identified through imagery and object detection models. Future physical risks, such as property loss due to wildfires, can be anticipated with historical data and predictive models. Investment risks, such as building retail sites in unprofitable locations, can be simulated by analyzing product sales and variables such as customer demographics.

These risk management practices rely on a form of AI called GeoAI. Modern GIS technology includes GeoAI, which detects and forecasts patterns in any kind of spatial data—imagery, weather, asset locations, buying patterns, and more.

Physical risk management, today and tomorrow

Physical risk management, such as repairing equipment or protecting facilities from storm damage, benefits immensely from GeoAI. Physical risks are inherently geospatial since each asset has a location. That location determines which risks the asset might encounter, the likelihood and extent of operational impacts, and the speed and cost of mitigation.

In one example, a global energy provider uses drones, imagery, and GeoAI to spot offline or damaged solar panels. With near real-time feedback about which panels need servicing, operations managers can calculate the cost of repairs, direct repair crews to the affected panels, and keep productivity loss to a minimum.

To manage future physical risks, Netherlands-based real estate firm Bouwinvest uses predictive models to forecast climate impacts to 2050. A GeoAI-backed climate dashboard shows the severity of impacts in each region where the company owns assets and drills down to show decision-makers whether individual properties are resilient against physical risks, such as waterlogging, wildfires, or extreme heat.

Seeing investment risks and opportunities

Predictive models such as those used by Bouwinvest also uncover investment risks. Understanding the likely impact of physical risk across a portfolio can shape decisions about mitigation as well as acquisitions and divestments. Financial leaders can use similar models to analyze whether loans might default because of climate risk, whereas a chief operations officer might spot partner companies whose supply chains are exposed to disaster-prone areas.

Managing investment risk could mean using GeoAI to model revenue at several proposed business locations. One retailer worked with a GeoAI expert to forecast its next decade of sales based on 10 years of transactions. Since those transactions occurred during an economic expansion, GeoAI simulated the risk of a downturn, giving executives a grounded view of each location's potential. The ability to answer key questions in advance reduces the risk of unprofitable outcomes, pointing decision-makers toward data-backed opportunities.

Understanding the when and where of risk management

On a chart or spreadsheet, leaders can see that risks exist—the dollar amount of potential losses or the percentage of risky assets in a portfolio. On a map created with GeoAI, leaders can see what these risks look like in the real world: where a hurricane could hit mortgaged properties or how supply chain shortages will cascade through a production network.

By pairing AI with GIS, leaders can understand the reality of shifting risks, seeing not only what's happening but what could happen and where. It's an application of innovative technologies that executives shouldn't ignore in the new era of AI-assisted business.

This story by Mansour Raad originally appeared as "Survey: CEOs May Be Underutilizing This AI Capability" in *WhereNext Magazine* on October 3, 2023.

Another AI Capability that Business Leaders May Be Overlooking

Esri

Executives might be overlooking some of AI's most valuable uses, according to a recent article in *Harvard Business Review* (*HBR*).

Although generative AI gets headlines for boosting productivity, analytical AI—including machine learning and deep learning—often delivers clearer business benefits, *HBR* reports. It's particularly effective for forecasting outcomes from the structured data that drives key business decisions, from sales figures to asset values to inventory levels. Leaders rely on such data to make decisions with bottom-line impact in areas from risk management and market expansion to capital investments.

What's more, many companies already have analytical AI in their existing systems. For example, GIS software uses GeoAI to map and analyze business data about store locations, supply chains, and logistics.

As businesses increase AI spending, *HBR* recommends using the two types together: analytical AI to process data and generative AI to make those insights accessible through natural conversation.

Analytical AI, generative AI: Getting the best of both

HBR describes how a major telecommunications company built a custom generative AI tool that answers questions about company data by automatically generating and running statistical analysis code.

Enterprise systems, such as GIS, could provide this capability without requiring custom solutions.

GIS has long excelled at harnessing GeoAI to answer critical business questions by examining large datasets, identifying trends, and modeling outcomes. For example, it can analyze historical climate data to predict weather patterns—crucial for understanding physical risk in industries such as insurance or utilities. Or it can produce sales forecasts by interpreting data such as storefront locations, customer transactions, and local demographics and psychographics, market research based on psychological variables.

Historically, running GeoAI analyses has required GIS or data science expertise. Visualizing the findings involved making maps—a professional specialty of its own. Now, GIS incorporates generative AI interfaces, allowing nontechnical users to query data and make effective maps, freeing up GIS professionals for more strategic projects.

Generative AI, *HBR* points out, is particularly useful for enhancing productivity in creative tasks. Through generative AI, users can ask for analyses and interact with the output, adding or removing datasets and emphasizing key findings, speeding up the time to insight.

Generative AI accelerates adoption and use of data

The *HBR* article notes that while analytical AI delivers measurable value, generative AI helps build support for it. Leaders interviewed by the authors said the attention on generative AI helped them gain backing for other AI projects, especially since nontechnical stakeholders now better understand AI's potential.

Ultimately, the two types of AI play complementary roles. Analytical AI works behind the scenes, whereas generative AI raises the technology's profile, powering interfaces that extend the value of AI analyses to more end users.

With both types of AI available in systems such as GIS, employees at all levels of the business can be empowered to make more informed decisions, share insights, and drive innovation.

This story by Jay Theodore originally appeared as "The AI Capability Business Leaders May Be Overlooking" in *WhereNext Magazine* on February 25, 2025.

Mapping New Possibilities for Business Success

Esri

Here's a simple reason why AI is the focus of so many conversations: With this technology, business teams can reset the boundaries of what's possible.

Consider a luxury seafood company based in Nova Scotia that has significantly increased its catch using AI-driven analysis of marine territories and species. Or the world's largest telecommunications company, which can determine where its communications towers and other equipment are most vulnerable to extreme storms and floods brought on by climate change. Similarly, a leading US shipping company uses AI to forecast when its cargo planes will require new parts—squeezing more efficiency out of its supply chains.

AI alone is transformative but its integration with the powerful mapping and data management tool of GIS, creating GeoAI, multiplies its impact. This powerful combination not only opens new doors but also provides a comprehensive toolkit for businesses to refine their strategies and operations. GeoAI enables a deeper, real-time understanding of business opportunities, environmental impacts, and operational risks, empowering companies to continuously innovate.

AI scours massive libraries of data to extract useful information. GIS organizes that information and makes it visible and ready for analysis on real-time maps and dashboards. This enables greater clarity for organizations to answer simple but critical questions including the following:

- Where exactly are the opportunities, impacts, and risks?
- Where are our best customers and locations, and where are they likely to be in the future?
- Where are essential resources, and how can we operate in those places with the least impact on the environment?
- Where are assets in danger from rising seas, extreme heat, or other climate risks?

With GeoAI, the answers are calculated at speeds and scales that were unimaginable only a few years ago. One executive has described GeoAI as "almost the holy grail for a maintenance operation."

Solutions designed for the unique challenges of the executive suite

Three powerful capabilities emerge with the joining of AI and GIS:

- Automate tasks and repeat them quickly at scale to optimize business processes across an enterprise while gaining situational awareness of operations, assets, and supply chains.
- Look at past patterns to make predictions and acquire insights for decisions based on predefined criteria or objectives.
- Find patterns hidden in large amounts of data to detect correlations within customer demographics, economics, and geographies.

The application of these capabilities is useful across a wide range of industries:

- A retailer considering where to locate stores or other physical assets can study available commercial properties, local customer preferences, and existing service providers.
- A business owner can define energy use patterns in buildings and pinpoint opportunities to lower costs and boost sustainability.

- A transportation planner can estimate how far people will likely travel for specific goods and services.
- An insurance company can predict flooding or wildfire damage—at the neighborhood level.
- A manufacturer or logistics team can optimize supply chains using disparate data, including weather forecasts, estimations of ship and rail traffic backups, or the number of left turns drivers make on their routes.

Additionally, GeoAI equips executives across an organization with a tool for problem-solving that uses the power of geography, or location:

- Chief executive officers can access near real-time analysis for the many decisions that hinge on where, when, and why. GeoAI is an enterprise tool with wide-ranging value in everything from planning how to drive growth to assessing the implications of policy changes and market trends.
- Chief operations officers can improve operational awareness across the enterprise with maps, dashboards, and remote sensing devices for monitoring assets and resources. They can quantify how events in one area impact the supply chain. Using GeoAI for prediction, they can prioritize asset and infrastructure maintenance, avoiding costly delays and shutdowns.
- Chief risk officers can map resources and assets to develop strategies for reducing exposure to climate impacts before they affect balance sheets. They can proactively reduce environmental impact. Also, forecasting to understand customer behavior across regions can guide decisions on spending and investment.

- Chief information officers can build technology solutions that advance their organization's operating model, improving service delivery and governance. Across the enterprise, business teams gain geographic context in their analysis of financial and customer relationship data. Off-site monitoring and data collection and management become more efficient and cost-effective with GeoAI tools for automation.

Predicting what's next

In an ever-changing business world, it's harder than ever to know what comes next. But with GeoAI, executives gain the data-driven insights and predictive abilities they need.

Across the enterprise, executives can use GeoAI to better visualize challenges and opportunities, deepen analysis, and deliver comprehensive strategies.

In short, they can redefine what's possible.

This story by Marianna Kantor originally appeared as "The AI Imperative—Mapping the New Possibilities for Business Success" at Forbes.com on September 20, 2024.

GeoAI, Reality Capture, and the Future of Digital Twins

Esri

One of our most persistent challenges in the facilities and indoor mapping domain has been helping customers quickly build adaptable systems of record that can serve as the foundation for a digital twin. The difficulty of creating reliable representations of the built environment on campus, in buildings, and in other indoor locations arises largely from traditional reliance on architecture, engineering, and construction (AEC) project documentation.

So, what's the problem?
- As-built conditions often don't match original plans.
- Building Information Modeling (BIM) models, although powerful, are often too complex for day-to-day operational use.
- CAD files are often poorly structured, while PDFs are unstructured, making integration into GIS difficult.

Paradigm shift: Reality capture and GeoAI

Our research at Esri has revealed a paradigm shift in how detailed and accurate building information can be generated through a combination of reality capture (RC) and GeoAI, enabling this data to readily provide customers with digital twin capabilities. In some respects, this feels like a "back to the future" moment.

Early in the author's career, in the Army and while managing the US Coast Guard's base mapping program, we relied on satellite and aerial data to capture reality snapshots. We then used a combination of manual digitization and early machine learning to extract planimetric features—roads, fence lines, and buildings. The accuracy was only as good as our sensors and ground control, but the fundamental principle was clear: The best source of truth is the real world.

Fast-forward to today, and we're witnessing a similar trend. Except now, instead of satellite imagery, we're using consumer-grade reality capture devices, including GoPro 360s, iPhones with lidar, and commercial drones.

The breakthrough: GeoAI-driven indoor mapping

Imagine sending a barely trained person to walk around a campus with a GoPro 360. Using GeoAI, that simple walk-through can generate high-fidelity 3D models of buildings and interiors. But it doesn't stop there—GeoAI can then automatically extract, classify, and structure the features inside those models into a spatially aware 3D GIS repository.

The floor plans extracted from these AI-driven models are often more accurate than computer-aided design (CAD) or building information modeling (BIM) files.

We validated this capability in a recent research project. A customer had previously captured lidar and panoramic imagery across millions of square feet of facilities to create BIM models. Later, when they required a comprehensive, spatially enabled security asset database, they'd encountered a lengthy manual effort to document every camera, sensor, and access control point.

Instead, we used GeoAI to analyze the existing data—extracting features from lidar depth and applying AI-based depth estimation to identify and classify security assets. The result? What would have taken years was automated in a fraction of the time.

Prediction and pattern recognition with GeoAI

Two of the most useful capabilities of GeoAI are predictive spatial analysis and pattern recognition—both critical to automating geospatial workflows.

In our Gaussian splat pipeline (a volume-rendering technique that deals with the direct rendering of volume data without converting the data into surface or line primitives), for example, we optimize radiance fields to predict and reconstruct missing frames, which allows us to fill in the gaps in captured data. This ensures smooth, accurate digital twins without requiring exhaustive, 100 percent coverage scanning. The same principles apply at larger scales—whether predicting infrastructure changes over time or reconstructing incomplete building models with AI-driven inference.

GeoAI is also fundamentally changing how we detect and extract key features. Rather than searching an entire dataset, we use models such as Contrastive Language-Image Pretraining (CLIP) to narrow the universe of images to search, prioritizing the most relevant data before extraction begins. Whether identifying specific building assets or classifying security infrastructure, this targeted approach reduces noise and increases accuracy, making feature extraction scalable and precise.

The new value proposition: GIS as the source of truth

All of this raises a fundamental question: Why not use GIS to update and create CAD and BIM rather than vice versa?

We're now in a world where it's easier than ever to do the following:
- Generate high-accuracy, photo-realistic 3D models of your buildings and interiors using Gaussian splatting and advanced GeoAI techniques.

- Extract the spatial features you need automatically using GeoAI-driven recognition and depth estimation.
- Turn those insights into a dynamic, operational GIS, serving as a system of record and a digital twin.

GeoAI saves time—it automates what used to take weeks or months. It enables living systems of record to continuously evolve with the real world rather than remain static representations.

The future of facility management: AI-powered digital twins

This convergence of AI, GIS, and reality capture fundamentally changes how organizations interact with their built environment. Instead of relying on outdated documentation, we can create spatially intelligent, continuously updated digital twins that reflect reality.

The implications are enormous:
- **Speed and scale:** Tasks that once required manual effort can now be automated at scale, unlocking new levels of efficiency.
- **Enhanced decision-making:** AI-powered insights enable faster, more informed decisions based on real-world conditions.
- **Collaboration and cloud integration:** GeoAI workflows support cloud-based collaboration and seamlessly integrate with industry-standard tools, such as NVIDIA Omniverse and other 3D ecosystems.

The big shift: From documentation to intelligence

A GeoAI-enabled GIS allows our users to leap beyond the foundational capabilities of a system of record, allowing them to create digital twins of their facilities and business operations. The future isn't just about visualizing facilities and business operations—it's about using the digital twin to understand and interact with them in real time to drive better outcomes in the real world.

The question isn't whether AI will transform geospatial workflows. It already has.

The real question is, how will you make the most of GeoAI and ArcGIS to redefine what's possible?

This story by Pat Wallis originally appeared in the *Esri Industry Blog* on February 24, 2025.

Mapping the What-Ifs: Fertile Ground for AI-Powered Simulations

Esri

Executives strive to make data-driven decisions whenever possible. But what if the data is too expensive, or altogether impossible to gather?

We synthesize it. The first computers were designed for this very purpose.

On the potentially disastrous Apollo 13 mission, NASA loaded telemetry data from the wounded spacecraft into a training simulator in Houston. There they devised maneuvers that brought the crew home safely.

Amazing what was achieved with a few bytes of data—the computing equivalent of a pocket calculator.

Today, we're working with powerful semiconductors and plentiful data. Already indispensable in science and industry, simulations have grown more practical, easier to visualize, and more accurate with advances in AI.

Putting these AI-powered simulations in a geographic context—tapping the power of location—offers tremendous business value.

Supply chain visibility and automation

One tech heavyweight has created a digital simulation to handle product and material requests for 130,000 customer locations around the globe. Its key customers can have the parts they need and a skilled engineer on their doorstep within two hours.

A geographic digital twin automates the hard part—making the right call about who to dispatch. Powering this digital twin, GIS technology combines data for inventory, customers, and service engineers with data used for routing and logistics.

Machine learning models within the digital twin are trained to find patterns in the data and predict the fastest delivery routes. Dashboards show the status of the work and illuminate opportunities for improvement.

Building climate-resilient infrastructure

Digital twins are great in a time crunch. These simulations are even better for visualizing longer-term projections.

Picture telecommunications infrastructure: a network of stations, towers, and cables spanning continents. Some of the most sophisticated machinery ever built exposed to the extremes of what Mother Nature can deliver.

Floods and high winds cost telecoms billions of dollars in infrastructure damage. This makes climate adaptation a smart business strategy.

The largest telecom in the world uses a digital twin to predict how weather conditions might affect its infrastructure—present day, in three years, and in 30 years. The organization interfaces with supercomputers at climate research institutes that generate scenario-based projections. Bringing in GIS data shows exactly where infrastructure might be at risk.

As climate dynamics change, so will the maps and simulations. And the climate prediction models will inform the placement of new cell towers and cable lines and show where to build mitigation measures around existing structures.

Public works in 7D

Other cutting-edge AI-powered simulations push the operational envelope into new dimensions.

A massive public rail project is under way in Brisbane as Australia's third-largest city prioritizes smart city infrastructure.

The project comprises two underground tubes and 1,300 kilometers of track tunneled through the heart of the business district to the suburbs. A dozen aboveground and underground stations will be built or rebuilt in already developed parts of the city. A tidal river prone to seasonal flooding adds a complication.

The design and execution of the project happens entirely inside a virtual simulation. Not just the stations, tracks, and tunnels; this digital twin captures 13,000 square kilometers of the city.

All crews and contractors were required to federate digital plans inside a common data environment. Data for construction, surrounding geography, drone imagery, lidar and sensor data, all combined in one place.

This created a platform for experiencing AI-powered simulations in lifelike detail. Machine learning algorithms in a video game engine render the imagery and complex 3D environments. Imagine Pokémon GO for engineers.

Within the simulation, designers experiment with aesthetic and functional details. Will the corridors have enough natural light? Will bicyclists collide with pedestrians? Should this public art exhibit face east?

Engineers consult virtual blueprints accurate to 2 millimeters and updated every four hours. 3D maps guide the work; the planners know the location of utility lines, sewers, and 200 other layers of geographic information.

Construction managers tap financial data to understand how much different scenarios will cost and affect timelines. The simulation tracks 200 million assets as they arrive at the jobsite.

In a 360-degree virtual reality theater, the city's train conductors gather to experience track lines before they are built so they can give feedback.

Construction engineers and government officials swap hard hats for VR headsets, taking a virtual stroll through a construction site.

Eventually, the digital twin will serve as an operational dashboard, providing real-time data and analytics to railway operators.

Moving from simulation to reality

In the GIS technology community, bridging the sim-to-real gap in this way has been a priority.

Purpose-built to analyze rich, structured, and diverse datasets, GIS technology is fertile ground for the integration and training of machine learning models. With the rapid expansion of enterprise data, we can address our toughest challenges with greater certainty.

Every scenario has clues that point to the best outcome. Great executives know how to identify the ideal scenario and bring about the best outcome. GIS technology has a way of making these patterns more apparent.

Today, GIS together with AI—GeoAI—empowers us to explore possibilities in ways that are smarter and more holistic than ever before.

A version of this story by Jay Theodore originally appeared as "Mapping the What-Ifs: Geospatial Technology Is Fertile Ground for AI-Powered Simulations" in *CIOReview* in March 2025.

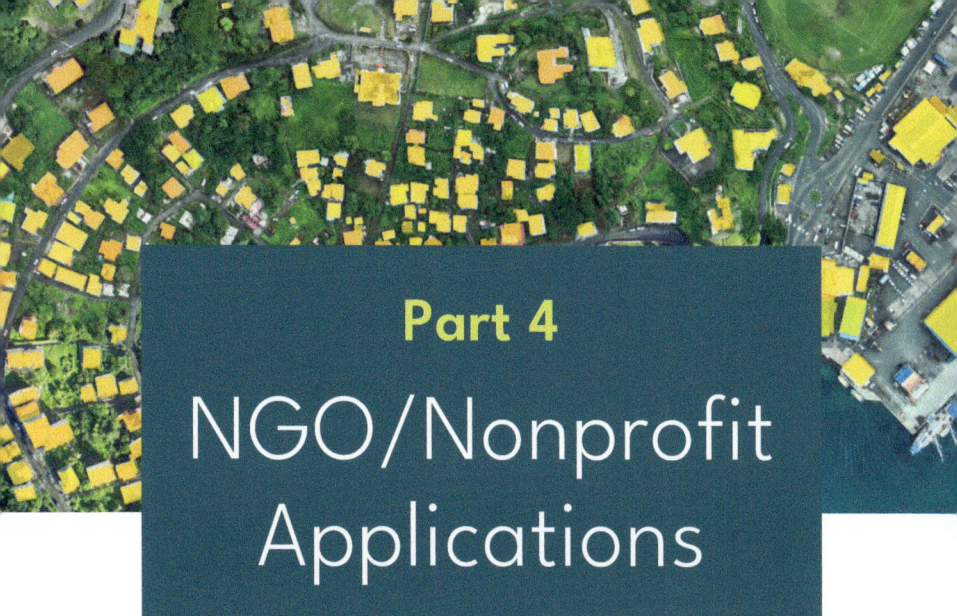

Part 4
NGO/Nonprofit Applications

In the NGO/nonprofit arena, even with limited technical staff, environmental organizations can better track deforestation, wildlife movements, and climate system dynamics using AI-powered pattern detection. Humanitarian organizations can use GeoAI to help them more quickly assess needs and coordinate aid during crises by asking questions about population displacement, resource distribution, and access to services.

GIS in action

This section presents real-life stories about how NGO and nonprofits use GeoAI to improve analysis, increase precision, accelerate action, and more.

Drone Mapping Helps Find Flood Victims, with AI Assistance

United Nations World Food Programme

Two back-to-back cyclones, Idai and Kenneth, battered Mozambique in 2019, destroying more than 800,000 hectares of farmland during harvest season. The devastation to crops and livelihoods left nearly two million people facing acute food insecurity.

The United Nations (UN) World Food Programme (WFP) responded quickly, with two helicopters ferrying supplies and rescuing stranded people. Given flooded roads, the air support was crucial but not nearly enough to distribute food and find stranded people across such a wide area of impact.

"When your primary tool for flying over large bodies of water is a helicopter that costs $2,800 an hour, you need to find a better way," said Patrick McKay, drone data operations manager for WFP.

Cyclone Idai was the first time WFP coordinated a fleet of drones for disaster response. The timing was right because WFP had conducted drone training with Mozambique's National Institute for Disaster Management and Risk Reduction (INGD) to build in-country drone pilot capacity. Also, Idai was an unusual storm in that it parked over land, dumped rain for days, and flooded an area spanning thousands of square kilometers.

"Each time the pilots came back, they told us the flooded area had gotten bigger," McKay said. "They'd go by a stadium, and

Mozambique's Institute of Disaster Management used drones to respond to Tropical Depression Desmond, which caused devastating flooding in Beira in January 2019. Photo courtesy of INGD/Antonio Beleza.

people were climbing higher and higher in the stands every time they flew past."

The drone pilots mapped the damage and searched for survivors, freeing up helicopter pilots to conduct rescue and supply missions. The images and videos the drones collected fueled updates to maps created with GIS technology that let all responders see what was going on.

Drone capacity building in Mozambique

Drones are now standard for every WFP emergency response, used early and often to find out how many buildings have been damaged and how far floodwater reaches, as well as to scout for survivors and map changing conditions.

WFP is usually among the first on the scene, because it has offices in more than 120 countries and leads clusters of UN emergency telecommunications and logistics staff who need to go in and set up before other teams arrive. WFP's use of drones helps UN Children's Fund (UNICEF) inspect schools, assists the World Health Organization (WHO) assess the status of clinics, and supports other UN organizations and in-country first responders rapidly understanding conditions on the ground. GIS brings together drone-collected data—with details about people and structures—in purpose-built maps and apps to meet each organization's mission.

"Before, only the people with access to a helicopter could see the damage and do assessments," McKay said. "But now with drone pilots mapping in really high detail, down to two-centimeter resolution, we could share that information with everyone and even do remote inspections."

WFP worked directly with Antonio Jose Beleza, deputy director of the National Emergency Operations Center (CENOE) at INGD. Eight of the pilots came from the Mozambique government, and with their local knowledge, they began mapping where they thought people were and still might be.

Every day, drone teams would meet in the morning, disperse, and meet in the evening to coordinate activities and process images to update maps. Evening meetings were for coordinating the next day's flights and the areas to cover. In the morning, the teams would assign people to ground vehicles, helicopters, and boats to reach flooded areas. The drone teams would check on villages, report if they were flooded or clear, and document the people walking through floodwater toward safety.

Automating the search for people surrounded by floodwater

Each drone flight can collect hundreds to thousands of images that need to be processed and reviewed to reveal the most urgent situations.

"The faster we can go through them, the faster we can get to the people that need our help," said Patrick McKay, drone data operations manager for the World Food Programme (WFP).

The cyclones in Mozambique in 2019 marked the first time WFP began using a model with AI machine learning algorithms to automatically find and classify damage to buildings, which eliminated the need for people to look at every image. But to locate people, a team of 20 people had to train the AI model over six weeks to achieve accurate results. This was not a workable solution considering the time-critical nature of search and rescue.

WFP then turned to Esri partner Synthetaic (now RAIC Labs), as part of a European Commission Humanitarian Aid (ECHOH)-funded wide-area search study, to solve the more challenging problem of using computer vision to find stranded people surrounded by floodwater. Synthetaic breaks images into tiles and employs an approach that doesn't require a pretrained AI model. With Synthetaic's workflow, WFP was able to achieve helpful results to find people in water in a few days.

"That's where our product RAIC [Rapid Automatic Image Categorization] comes in," said Corey Jaskolski, president and founder of Synthetaic. "We can take completely unlabeled data, we don't need any humans to label it, and we can search for things in the dataset by a single example query."

Aided by AI in this way, responders could quickly send a boat or helicopter if the algorithm found a person in a tree surrounded by floodwater.

The mapping effort to prepare and respond

In March 2019, Cyclone Idai displaced hundreds of thousands of people in Mozambique. Six weeks later, Cyclone Kenneth brought powerful winds and heavy rains to the same devastated area. The

The drone team looks at an orthophoto to prepare data capture missions along the Buzi River. Photo courtesy of INGD/Antonio Beleza.

quick revisit was the first time in recorded history that two strong tropical cyclones hit the same country in the same season.

Flooding in Mozambique had started even earlier, in January 2019, with Tropical Depression Desmond, which dropped more than 400 millimeters (15¾ inches) of precipitation in less than 48 hours in Beira, the country's second-largest city. The entire city was flooded, which led Beleza to try drones to assess the damage. "At that time, it was just me and my colleague Agnaldo Bila with two drones," he said. "We couldn't map very large areas, but we were able to cover critical areas and share the footage in real time. The images we collected on the ground were integrated in real time and for the very first time into the European Union's Copernicus Emergency Mapping Service, which was activated at the request of WFP to conduct rapid assessment of flooded areas." The map helped establish accommodation centers in places that weren't flooded and showed others in the government the value of drones and mapping.

"Initially, we were focused on disaster response, but we wanted to be proactive," Beleza said. "Disasters will occur, and we want to

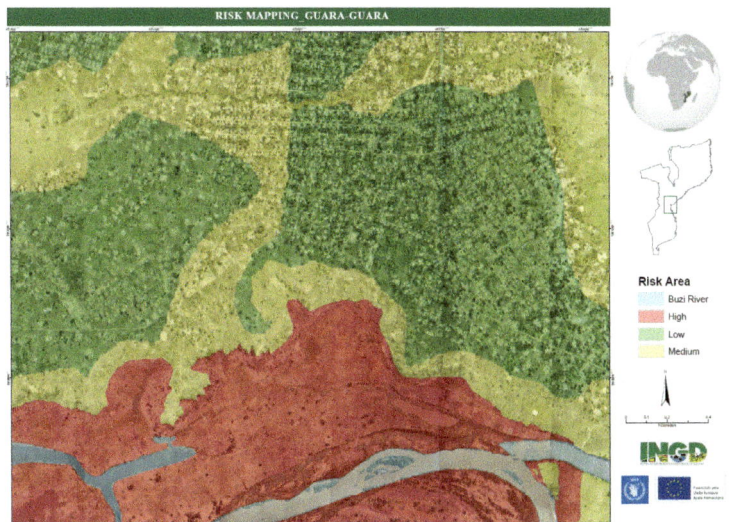

With the detailed DTM of the Buzi River, hazard maps were created to show the areas most vulnerable to flooding. Red areas show high vulnerability, yellow show medium, and green show low. Map courtesy of INGD/Antonio Beleza.

be more prepared. We don't want extreme events to turn into disasters anymore."

Beleza and his team have been working with the Italian Centro Internazionale in Monitoraggio Ambientale (CIMA Foundation) on hazard mapping. The CIMA Foundation has created a hydrological model for the Buzi River watershed using satellite imagery, and recently drones were flown to improve the accuracy of the model.

"We flew the drones over 850 square kilometers of the Buzi River, then we processed the data and created a very high-resolution digital terrain model (DTM)," Beleza said. "When the rainy season comes, local governments can simulate floods and determine when and where the water will arrive downstream. We're sure that this will trigger community early actions and save lives."

GIS is used to look at different layers of data to model flooding

WFP conducted drone training in Addis Ababa, Ethiopia, in April 2019.
Photo courtesy of WFP/Katarzyna Chojnacka.

scenarios and design evacuation routes and identify the safest places for accommodation centers.

"We combine science, technology, and local knowledge to prepare the communities and local governments in a participatory manner," Beleza said. "Every year we are seeing more frequent flooding. That has motivated us to go there and do something for these people."

WFP is greatly encouraged with how colleagues in Mozambique continue to grow their skills.

"They've got the expertise and drone equipment, so when the storms hit this year, they didn't call us because they knew exactly what needed to be done," McKay said.

For Beleza's team in Mozambique, the drones and AI analysis have been a game changer.

"We've been saving lives," Beleza said. "We are taking the opportunity to learn from the drone deployments we have made so far to improve the way we deal with this technology when it comes to disaster preparedness and response."

World Food Programme's expanding drone vision

WFP has conducted drone capacity training around the world, including in Bolivia, Colombia, the Dominican Republic, Ethiopia, Guatemala, Myanmar, Nepal, Niger, and Peru. With the success of Mozambique, WFP is busy adding capacity in the nearby countries of Madagascar, Comoros, Malawi, Lesotho, and Eswatini, where repeat flooding is a problem as well.

WFP's drone strategy has evolved over time.

"With drones, you either scale up the drone so you get bigger, more expensive, more capable drones, or you scale up the number of drones," said Patrick McKay, drone data operations manager for WFP. "Lately, we found that scaling the number of drones is more effective than scaling the capability of the drone."

A recent potential addition to WFP's drone kit is a prototype tethered drone to deliver a Wi-Fi signal across an area of several kilometers. With this connectivity, the data captured by drone pilots could be uploaded after every flight, feeding maps in near real time and reducing the reaction time to help people caught in floodwater.

WFP continues to try new approaches and test scenarios to improve humanitarian missions.

"I think the future is drone swarms," McKay said. "I want to put a laptop on the ground, add all drones, put landing pads around us, and press a button. The drones would then fly out in a pizza pie pattern, where each goes and does its slice of the pie and comes back home."

WFP is looking into contracting cargo drone services to carry out deliveries ranging from very small medical packages—under five kilograms—to multiple tons of food. This could help reduce the cost of airborne cargo delivery and allow WFP to deliver within conflict zones without putting pilots at risk.

Drones proved to be a powerful tool for WFP to deal with complex geography and reach places that lack road infrastructure. In Africa, where roads typically are washed away each rainy season, distances can be small, but the travel time can be large.

"That's the beauty of drones," McKay said. "It doesn't matter what's underneath it—as long as there aren't high winds or pouring rain, the drone can get there."

This story by Olivier Cottray originally appeared as "Drone Mapping in Mozambique Helps Find Flood Victims, with AI Assistance" in the *ArcGIS Blog* on November 10, 2022.

GeoAI, Corporate Responsibility, and the Vigilance of a Climate Watchdog

Amazon Conservation

There's a new breed of environmental watchdog out there. A decade ago, their predecessors walked through jungles and woods to document deforestation from illegal mining, oil drilling, cattle ranching, and palm oil farming. What then took months or even years to uncover, today's watchdogs reveal in a day, using satellite imagery and AI-powered GIS technology.

Dr. Matt Finer, senior research specialist at Amazon Conservation Association, helped usher such techniques into mainstream use. Today, as he and his colleagues perform real-time monitoring and analysis to protect productive land and stave off the effects of the climate crisis, they're infusing the role of global watchdog with even more speed and precision.

A watchdog peers deep into the supply chain

For global businesses, total supply chain visibility is rare—some studies say just 6 percent of companies have achieved it. But executives can no longer plead ignorance of the activities occurring deep in their supply networks. New and pending legislation cracking down on deforestation, for instance, proves that the actions of any player in the supply chain—however minor or far from headquarters—now pose a corporate risk.

And the watchdogs are watching.

Matt Finer's name is well known among them. In 2013, he took a job at Amazon Conservation, a nonprofit working to conserve the biodiversity of the Amazon basin; two years later, he founded the organization's Monitoring of the Amazon Project (MAAP) with a mission to fine-tune and fast-track the monitoring of conservation land. Instead of relying on field crews, Finer and the MAAP team use satellite imagery and GIS analysis to spot illegal deforestation. In the early days of the project, they created yearly reports based on 1,000-meter-resolution imagery. Now, they spot encroachments in areas smaller than half a tennis court—within a day or two of their occurrence.

Finer gets as many as a thousand alerts daily indicating that a given pixel in satellite imagery has changed color since the flyover a day earlier. The change could indicate that a few trees may have been felled or that smoke might be rising. GIS software spares the MAAP team from manually investigating each alert. Powered by data science, GIS groups pixels into clusters of activity, guiding MAAP and Amazon Conservation researchers to the most urgent areas of deforestation.

This watchdog work is happening at a time when the Amazon is in danger of losing its status as the planet's carbon sink. Scientists say tree loss and fires threaten to turn the Amazon into a net emitter of carbon dioxide, which would hasten atmospheric warming and worsen climate change.

Finer and his team watch for activities that upset the natural balance. With the evidence they've compiled, authorities throughout the Amazon have shut down activities ranging from illegal palm oil production to destructive cattle ranching in endangered areas to illicit gold mining.

New maps empower the protectors

Nadia Mamani was born and raised in the biodiverse Madre de Dios region of Peru, where gold miners and others have illegally cleared the rain forest to set up operations.

Despite having lived in Madre de Dios all her life, she hadn't grasped the extent of the illegal mining until she began using GIS and remote sensing tools to complement her field investigations. What she saw inspired her to work for conservation and restoration of landscapes in Indigenous communities.

That newfound awareness parallels what a business executive might see when GIS-based insight reveals unwanted activity deep in the supply chain, otherwise out of sight.

"In my hometown, there are some areas that are not very accessible," Mamani explains. "So, for me, GIS and remote sensing gave me new eyes—a view from a different angle."

During her graduate studies in GIS and conservation, Mamani relied on Finer's maps to investigate mining activity in her town. "Matt Finer is a household name in the conservation community in Peru," she explains.

Upon moving to the US to start her career, she had a professional goal.

"I want to find Matt Finer," Mamani told colleagues.

When she found him at Amazon Conservation in Washington, DC, she also found work as a GIS and remote sensing specialist for the MAAP team.

"After I moved here, I decided to work for my country and my hometown. And in some way, that's what we do—protecting my people, our forest. It's my home, and that's something that unifies me and Matt. He's always [taking] scientific data, information—basically complex problems—and turning them into digestible public reports. That's something that he does perfectly. For me, it's translating complex ideas into something simple."

A pioneer in the right place at the right time

In the middle of the last decade, Finer and the MAAP program scored a major win by uncovering a massive deforestation project—2,000 hectares cleared in a couple of months by a cacao producer.

The team had used GIS software to combine satellite imagery with data on the location of roads, country boundaries, mining concessions, and protected land.

"With some concise text, [we] kind of walked the reader through this case. And you may not realize it, but we just brought together 10 data layers," Finer says of the analysis. "It's really in these GIS programs ... where all of that data comes together."

That method showed its effectiveness when *El Comercio*, Peru's oldest and most popular newspaper, printed the satellite images of the cacao operation on its front page. When authorities broke up the operation, *El Comercio* printed that news, too.

"It was really just a game-changer moment," Finer recalls.

Cat and mouse in the jungle

But while watchdog groups such as Amazon Conservation were using technology to detect and expose some threats, others were going undetected.

Sidney Novoa, director of GIS and technology at Conservación Amazónica, Amazon Conservation's sister organization in Peru, watched the process unfold. His early work focused on illegal timber operators who built huge roads and extracted vast swaths of wood from the rain forest. Smaller operations that worked as discreetly as possible, using small trucks and removing only select trees, often escaped notice.

The ever-increasing clarity of satellite images and the emergence of GeoAI soon gave the MAAP team a new advantage. When loggers try to thin trees in small numbers, the activity creates subtle but telltale gaps in the canopy.

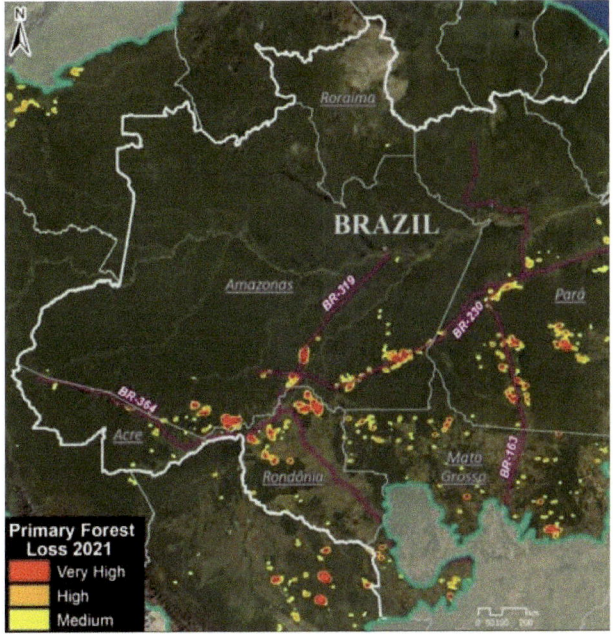

A GIS-based map produced by Amazon Conservation shows areas of concern.

"And that characteristic can be identified by [machine learning] algorithms once you've trained the [GIS program on the] shape of the gap and, sometimes, the color of the area that the operation leaves," Novoa said. "We identified camps, we identified the timber wood ready to be transported up to other areas. We reported that to the local authorities here in Peru."

In other words, GeoAI reveals what some people hope will remain hidden.

A monthlong trip becomes a career

Finer never expected to spend his career safeguarding the Amazon, but from the start, he was drawn to expansive landscapes and complex challenges. Shortly after completing his PhD at Washington

State University, he took a trip to Ecuador. He had a ticket to return in a month, but he stayed a year, and when he left, he was well on his way to a career of Amazon watching.

As for what motivated him, he says, "I think it was a combination of the biodiversity [and] the vastness that was still intact but combined with the threats."

His brush with satellite imagery and GIS technology—the tools that helped him and Amazon Conservation define the practice of real-time monitoring—was just a happy accident.

"I have no background or training in technology whatsoever," he says. "During that time period, I organically discovered that technology was the most powerful tool to basically document in a scientific way ... what was happening."

As pressure on the business world grows, more corporate executives will rely on watchdogs like Finer—or location analysts within their own organizations—to root out bad practices from supply chains.

Although Finer and his MAAP colleagues address an array of threats across the Amazon, Peruvian gold mining helps illustrate the team's impact and the power of watchdogs to change business practices.

"We've really been able to track that problem in real time, send that information to the Peruvian government, and really help the Peruvian government clamp down," Finer explains. "And we've seen a major decrease in illegal gold mining deforestation in the Peruvian Amazon. So, it's really helping propel this field of real-time monitoring."

This story by Alexander Martonik originally appeared in *WhereNext Magazine* on May 17, 2022.

Mapping Land Mines and Explosive Remnants of War

The HALO Trust

Ever since the war in Ukraine began, humanitarian assistance organizations have gathered in Rzeszow, Poland, to aid displaced people and rebuild when the fighting ends.

Many have been mapping the damage inside Ukraine to prioritize the reconstruction and resettlement necessary to safely return Ukrainians to their homes. In just the first month of the conflict, more than 4 million people fled the country, and an estimated 6.5 million people were internally displaced.

For The HALO Trust (HALO), the largest demining organization in the world, the mapping effort is focused on identifying the presence of explosive remnants of war as well as damage to homes and infrastructure.

A geographically dispersed team is using a GIS to detail the impacts and the ongoing dangers.

"What we do is critical to reconstruction and resettlement post-conflict, because you cannot just deal with the damage straight away," said Luan Jaupi, head of information and communications technology at The HALO Trust. "We enable other humanitarian organizations and national authorities to safely conduct their activities by informing them where it's safe to go and making places safe by clearing the explosives that are littered around."

Accessing open-source information

Social media and commercial satellite imagery have allowed the world to see and document the war in Ukraine. These so-called open-source intelligence sources differ from classified intelligence that can be shared only through diplomatic channels.

"We're collecting information from internet sources such as Twitter [now X], Telegram, and Facebook," said Jesse Hamlin, global GIS and database officer at The HALO Trust, in 2022. "We're finding lots of tanks and armored personnel carriers that have been destroyed, and they're potentially booby-trapped with land mines around them. We're also seeing mines being littered across a road and vehicles just driving past them, because there's a panic to get out."

HALO's explosive ordnance experts review a variety of datasets flowing into its database from social media and news outlets, verify whether the data is relevant to the mine action sector, identify the model of the munitions if possible, and then place the dangers on a smart map to be shared with others.

Much of the destruction in Ukraine surrounds civilian areas, such as this damaged apartment building in Kyiv.

"We're finding lots of bridges that have been blown up by both sides, which means you may have been able to cross that bridge at one point, either to leave Ukraine or to come back in after the conflict is finished, but now you physically can't cross the bridge until it is repaired," Hamlin said. "This will impact the delivery of aid and the return of the people to their communities."

Putting evidence on the map

The HALO team has worked to streamline the steps of geolocating an event on social media and to speed the flow of information. Experts can now use geospatial technology to go through a stream of evidence and filter out just the events they feel need to be investigated, rather than manually searching the internet for social media and news articles.

"Looking at the map, the areas in red show the line of contact, where Russians are pushing forward and getting pushed back on a daily basis," Hamlin said. "The dots show events we pulled from

Local unexploded ordnance experts deal with rocket remnants in Mykolaiv, Ukraine. Photo courtesy of Anastasia Prokofyeva.

social media that are color coded for unexploded ordnance, land mine, improvised explosive device, cluster munitions, and other bomb types."

Each social media post is explored by HALO's explosive ordnance disposal experts, who examine the image to see what they will be dealing with. "They know right away if it's a FAB-500 aircraft bomb, which provides an evidence point we can investigate in the future," Hamlin said.

Mapping the bombs and bomb types has helped HALO educate the public in Ukraine about the munitions being dropped.

"We're interested in cluster munitions for one, because they're dispersed over large areas and children often pick them up and play with them," Hamlin said.

HALO has also received evidence of a new type of land mine called the POM-3, with sensors that detect human footfall rather than being triggered when disturbed. This new type of mine is launched by rocket and falls to the ground by parachute. When it senses a person, it detonates an explosive that spreads fragments in a circle 50 yards across.

The POM-3 adds great complexity and danger to the demining effort because it will require bomb squad robots to dismantle these mines at a distance, and HALO will need to acquire that technology.

Automating damage detection

HALO first did this advanced mapping work in Tripoli, Libya, in 2019, for a conflict that lasted a year.

"We asked ourselves what we could be doing during the conflict to aid our postconflict intervention, and we decided to record and map the presence of explosive ordnance," Jaupi said. "We learned a lot of lessons, and we're doing things a lot better now."

In some of the cities in Ukraine, such as Mariupol, early imagery showed massive damage from excessive and constant shelling. An effort is underway to apply the AI approach of machine learning to train computers to detect the damage from the imagery. HALO has been working for some time with machine learning experts at Esri to analyze the damage in countries such as Sri Lanka, Afghanistan, and Libya. Being able to detect damage programmatically helps HALO understand where to find unexploded bombs and land mines.

Work is still in progress to refine the dataset and maps that will be shared with other humanitarian organizations and the international community. As HALO continues to process and refine the dataset, using machine learning, its work will provide more clarity on the contamination inside Ukraine.

Returning to a war-torn land

HALO has more than 8,000 explosive ordnance disposal experts operating in 28 countries. It has been active in the Donbas region of Ukraine since 2016, removing explosive remnants of war from the 2014 conflict. And now there's a much larger job ahead because the conflict spans much of the country—the second largest in Europe, behind Russia.

When the situation begins to de-escalate, HALO will use the map to prioritize its work and keep its own staff safe. A simple GIS-based form is being developed that Ukrainians will be able to use to report what they find and where it is located so that experts can follow up and dispose of it. And HALO mappers plan to continue monitoring social media.

"Our goal is to ensure families can return and rebuild their lives in safety, and maps help us do this," Hamlin said.

This story by Ryan Lanclos and Olivier Cottray originally appeared as "Ukraine: The HALO Trust Maps Landmines and Explosive Remnants of War" in the *ArcGIS Blog* on April 12, 2022.

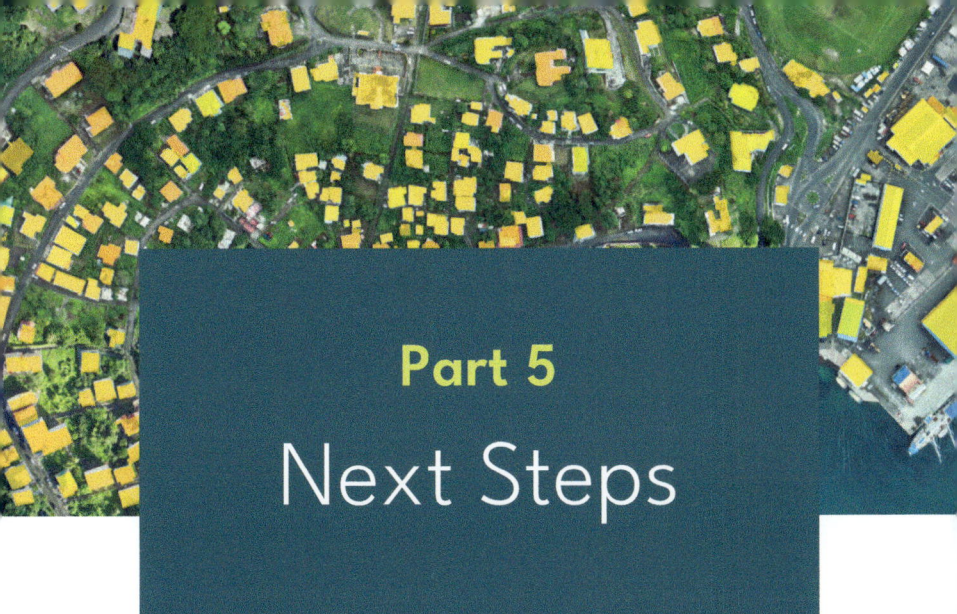

Part 5

Next Steps

GIS technology has long helped leaders solve problems by analyzing data through the lens of location, and AI makes GIS easier to use and even more powerful. Here are some ways you can get started using GeoAI to make better decisions faster.

Machine Learning with ArcGIS Pro

Machine learning refers to a set of data-driven algorithms and techniques that automate prediction, classification, and clustering of data. Machine learning can play a critical role in spatial problem solving in a wide range of application areas, from image classification to spatial pattern detection to multivariate prediction.

In addition to traditional machine learning techniques, ArcGIS also has a subset of machine learning techniques that are inherently spatial. These spatial methods, which incorporate some notion of geography directly into their computation, can lead to deeper understanding. The spatial component often takes the form of some measure of shape, density, contiguity, spatial distribution, or proximity.

Both traditional and inherently spatial machine learning can play an important role in solving spatial problems, and ArcGIS supports their use in several ways.

Machine learning can be computationally intensive and often involves large and complex data. Esri's continued advancements in data storage and parallel and distributed computing make solving problems at the intersection of machine learning and GIS increasingly possible.

Learn more:
- "Predict Seagrass Habitats with Machine Learning"
  go.esri.com/geoai_book1
- "Discovering Alternate Climate Zones Through Machine Learning"
  go.esri.com/geoai_book2
- "Build House-Valuation Models with Machine Learning"
  go.esri.com/geoai_book3

Deep Learning with ArcGIS Pro

Remotely sensed imagery can contain a wealth of information, from the number of buildings in a city to the type of crops being grown in fields across the world. Historically, to extract the buildings or swimming pools or palm trees in an image, you would have needed to manually digitize each feature, a process that could take weeks or years depending on the size of the image. But with improvements in computing power and new, accessible tools for deep learning in ArcGIS Pro, it's not difficult to train a computer to do the work of identifying and extracting features from imagery.

At the highest level, deep learning, which is a type of machine learning, is a process in which the user creates training samples—for example, by drawing polygons over rooftops—and the computer model learns from these training samples and scans the rest of the image to identify similar features.

Learn more:
- "Deep Learning with ArcGIS Pro Tips & Tricks: Part 1"
  go.esri.com/geoai_book4
- "Deep Learning with ArcGIS Pro Tips & Tricks: Part 2"
  go.esri.com/geoai_book5
- "Deep Learning with ArcGIS Pro Part 3: QA/QC Extracted Features"
  go.esri.com/geoai_book6

Sample Notebooks

Sample notebooks demonstrate various features of ArcGIS API for Python. The samples are categorized by the user profile they are most relevant to. Most samples are in the form of a Jupyter Notebook that can be viewed online or downloaded and run interactively. A few samples are provided as stand-alone Python scripts in the GitHub SDK repository.

Learn more:
- Sample Notebooks | ArcGIS API for Python
  go.esri.com/geoai_book7

Getting Started with ArcGIS Pretrained Models

ArcGIS pretrained models automate the task of digitizing and extracting geographic features from imagery and point cloud datasets.

Manually extracting features from raw data, such as digitizing footprints or generating land-cover maps, is time consuming. Deep learning automates the process and minimizes the manual interaction necessary to complete these tasks. However, training a deep learning model can be complicated, as it needs large quantities of data, computing resources, and knowledge of how deep learning works.

With ArcGIS pretrained models, you do not need to invest time and effort into training a deep learning model. The ArcGIS models have been trained on data from a variety of geographies and work well across them. As new imagery becomes available to you, you can extract features and produce layers of GIS datasets for mapping, visualization, and analysis. The pretrained models are available in ArcGIS Living Atlas of the World to anyone with an ArcGIS account.

Learn more:
- "ArcGIS Pretrained Models"
  go.esri.com/geoai_book8

Free Hands-On Learning

Hands-on learning can strengthen your understanding of GeoAI and how it can be used across sectors. Esri provides a collection of free story-driven tutorials that allow you to experience GIS applied to real-life problems (see the link below):

- **"Get Ready for Deep Learning in ArcGIS Pro"**
  Install the deep learning libraries needed to run deep learning workflows in ArcGIS Pro and learn to troubleshoot the most common issues.
- **"Extract High-Resolution Land Cover with GeoAI"**
  Use a deep learning pretrained model to extract land cover from high-resolution drone imagery.
- **"Train a Model Using Automated Deep Learning"**
  Use the Train Using AutoDL tool to train several deep learning models and pick the best-performing one for a pixel-level land-cover classification task.
- **"Map Floods with SAR Data and Deep Learning"**
  Use a deep learning pretrained model to extract water pixels from pre- and postflood Sentinel-1 datasets and perform change detection analysis to identify flooded areas in the St. Louis, Missouri, region in 2019.
- **"Detect Objects with Text SAM"**
  Use a multipurpose GeoAI model with a free-form language prompt to detect boats in Copenhagen imagery.

- **"Detect Objects with a Deep Learning Pretrained Model"**
  Use a GeoAI tool and a pretrained model to automate palm tree detection.
- **"Improve a Deep Learning Model with Transfer Learning"**
  Use transfer learning to fine-tune a deep learning pretrained model in ArcGIS Pro and obtain enhanced results when extracting building footprints in a Seattle neighborhood.
- **"Identify Infrastructure at Risk of Landslides"**
  Extract building footprints from imagery using deep learning and apply raster functions to perform a landslide susceptibility analysis.
- **"Extract Informal Settlements with SAMLoRA"**
  Train a SAMLoRA (Segment Anything Model with Low-Ranking Adaptation) deep learning model in ArcGIS Pro to extract informal settlement building footprints from drone imagery.
- **"Classify Power Lines Using Deep Learning"**
  Perform lidar point cloud classification using deep learning techniques to classify power lines.
- **"Train a Model to Identify Street Signs"**
  Build and verify a model that can be used to automatically identify street signs with ArcGIS Survey123.
- **"Classify Mangroves Using Deep Learning"**
  Use deep learning to determine the extent of mangrove forests in Mumbai, India, and how their footprints have changed over time.

Try these free GeoAI tutorials and more at: **go.esri.com/geoai_book9**.

Learn More

Learn more about GeoAI by visiting: **go.esri.com/geoai_book**.

Contributors

Rami Alouta

Matt Ball

Jim Baumann

Linda Beal

Clint Brown

Chris Chiappinelli

Jack Dangermond

Anastassios Dardas

Rob Elkins

Genevieve George

Kate Hess

Rohit Singh

Alex Smith

Ben Smith

Citabria Stevens

Vinay Viswambharan

Robert Waterman

Jessica Wyland

About Esri Press

Esri Press is an American book publisher and part of Esri, the global leader in geographic information system (GIS) software, location intelligence, and mapping. Since 1969, Esri has supported customers with geographic science and geospatial analytics, what we call The Science of Where. We take a geographic approach to problem-solving, brought to life by modern GIS technology, and are committed to using science and technology to build a sustainable world.

At Esri Press, our mission is to inform, inspire, and teach professionals, students, educators, and the public about GIS by developing print and digital publications. Our goal is to increase the adoption of ArcGIS and to support the vision and brand of Esri. We strive to be the leader in publishing great GIS books, and we are dedicated to improving the work and lives of our global community of users, authors, and colleagues.

Acquisitions

Stacy Krieg
Claudia Naber
Alycia Tornetta
Jenefer Shute

Product Engineering

Craig Carpenter
Maryam Mafuri

Editorial

Carolyn Schatz
Mark Henry
David Oberman

Production

Monica McGregor
Victoria Roberts

Sales & Marketing

Eric Kettunen
Sasha Gallardo
Beth Bauler

Contributors

Christian Harder
Matt Artz

Business

Catherine Ortiz
Jon Carter
Jason Childs

Related titles

The Power of Where

Jack Dangermond, Esri, and the GIS Community

9781589486065

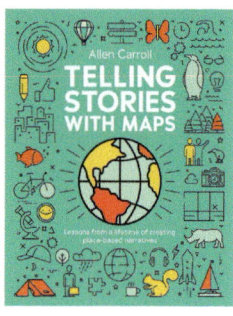

Telling Stories with Maps

Allen Carroll

9781589487970

Esri Advanced Guide to Python in ArcGIS

Dave Crawford & Daniel Yaw

9781589488236

Getting to Know Web GIS, fifth edition

Pinde Fu

9781589487277

For more information about Esri Press books and resources, or to sign up for our newsletter, visit

esripress.com.